STUFF YOU DON'T LEARN IN ENGINEERING SCHOOL

STUFF YOU DON'T LEARN IN ENGINEERING SCHOOL

Skills for Success in the Real World

CARL SELINGER

IEEE Press

A JOHN WILEY & SONS, INC., PUBLICATION

For general information on our other products and services please contact our Customer Care Department within the U.S. at 877-762-2974, outside the U.S. at 317-572-3993 or fax 317-572-4002.

Wiley also publishes its books in a variety of electronic formats. Some content that appears in print, however, may not be available in electronic format.

Library of Congress Cataloging-in-Publication Data is available.

ISBN 0-471-65576-7

Printed in the United States of America.

20 19 18 17 16 15 14 13 12 11

*To my wife Barbara, for her love and support,
and to my children, as well as to all the young engineers
who have inspired me to help them to be more effective
professionals and happier men and women.*

CONTENTS

PREFACE

I usually begin my seminar, "Stuff You Don't Learn in Engineering School," by saying that I wished I had attended a seminar like this when still a young engineer. Learning these nontechnical "soft" skills would undoubtedly have changed the trajectory of my professional and personal lives in very positive ways. As you'll soon see, for example, I would not have been afraid of "negotiating" for more than 20 years nor have worried about other skills that I never learned in engineering school.

So imagine my surprise and pleasure to see for the first time while I was writing this book that Dale Carnegie, way back in 1936, in his preface to his now-classic book *How to Win Friends and Influence People,* said "How I wish a book such as this had been placed in my hands twenty years ago! What a priceless boon it would have been." Carnegie went on to assert that his book was not just for those in business, but that *engineers* also needed these skills in dealing with people. He asserted that about 15% of one's financial success is due to one's technical knowledge and about 85% is due to skill in what he called "human engineering"—personality and the ability to lead people.

Carnegie went on further to describe the courses he conducted among engineers, including the New York Chapter of the American Institute of Electrical Engineers. Imagine, then, my surprise to note that, some 70 years later, I find myself walking down the same path to the publication of the book you're about to read, following *my* seminars to the New York chapter of the Institute of Electrical and Electronic Engineers (IEEE) and the subsequent support of IEEE Press in publishing this book!

I've written this book to give young engineers a practical, down-to-earth guidebook to the "real world" they are in, a very different place than the strenuous "boot camp" of engineering school.

Think of it as a "cheat sheet" or "crib sheet" for your current life, with tips and suggestions covering the most important nontechnical skills you need to learn to be effective and happy in your work and personal life.

Why is this so important? I agree with Dale Carnegie that the person who has the technical knowledge *plus* the ability to express ideas, to assume leadership, and to arouse enthusiasm among people, will be successful. Bringing this up to the present time, a Penn State professor recently told me that employers hiring young engineers *expect* them to have the technical know-how, but use their abilities in the nontechnical "soft" skills as *discriminators* in hiring them.

I do not subscribe to the old adage that "what you don't know won't hurt you." I believe that what you don't know *will* hurt you, or at least hold you back. You need to learn these important skills and achieve a proficiency in them if you want to be the best you can be, to reach the extent of your abilities. Every young engineer must know how to speak in front of a group, must be able to run a meeting effectively, must be able to understand and work with all sorts of people, and so forth. The good news is that you can *learn* these skills. If I can learn them, then you can learn them. This book is designed to explain over a dozen different skill areas in ways that you can understand, and tips that you can apply immediately to your situation.

My sincere hope is that this book will change the way that young engineers can become more effective and happier people in our complex real world. I hope this is what you find, and look forward to hearing from you about your successes. Good luck!

CARL SELINGER

Bloomfield, New Jersey
April 2004

CHAPTER 1

STUFF YOU DON'T LEARN IN ENGINEERING SCHOOL

What you don't know *will* hurt you and hold you back.
—Consulting engineer at Cooper Union seminar

I CLEARLY REMEMBER the date when *Stuff you Don't Learn in Engineering School* was, shall we say, "born." I remember this clearly because I was about to confront something in my career after 20 years of stressing over it. It was the morning of Thursday, October 20, 1992, around 9 AM, on the 61st floor of One World Trade Center, in New York City. I was sitting with about 20 other colleagues from The Port Authority of New York and New Jersey, and we were going to take a two-day course in "Negotiating Skills."

Why was I taking a "negotiating" course? After all, I was a seasoned manager in my forties, with over 20 years' experience, and yet I felt that I finally needed to learn this skill since I was about to negotiate with several bus and limousine companies to start ground transportation services to our regional airports.

You see, I had never learned how to negotiate. To be honest, I had a "thing" about it, and was scared of negotiating: I thought it was too hard, it always involved conflict, and you either won or you lost. As a result of these perceptions, I had avoided negotiating like the plague during my career. So now it was time to give in and finally learn how to do this terrible thing.

People were introducing themselves—who they were, what they did, why they were there, the usual. Then the instructor—a lawyer, whose name I've forgotten—started off the course. "Welcome," he said, "glad you're here," . . . then, "Why do you think they hired a lawyer to teach you how to negotiate?" My ears perked up. "We're the world's worst negotiators."

Stuff You Don't Learn in Engineering School. By Carl Selinger.
ISBN 0-471-65576-7 © 2004 the Institute of Electrical and Electronics Engineers, Inc.

Inside my head I was suddenly tuning him out, thinking, "this guy's a jerk." He continued: "Everyone thinks lawyers know how to negotiate. What we do is get together for lunch, and I say 'We need to settle this for $40,000,' and the other lawyer says, 'No, I'll only go for $30,000,' and, guess what? We 'split the difference,' settle for $35,000, and bill the client for our services."

Just that quickly, I had it with Mr. Negotiating Attorney! I didn't need to hear this bull for two whole days. I was going to call my secretary at the first break so she could come and retrieve me for an "important meeting."

The lawyer continued: "That's what most people think negotiating is. You say a number, I say a number, and we 'split the difference.' Well, I'm here to tell you that you already know how to negotiate." What was he saying?

"You negotiate every day. You've negotiated since you were a kid, asking to stay up later, wanting to borrow the family car. You negotiated all the time in college, begging the professor for more time to do the assignment. And you negotiate every day at work, asking for extensions for due dates, discussing changes in tasks you're supposed to do, selecting places to go for lunch, etc., etc. You just don't call it 'negotiating'."

Suddenly, something went off in my head . . . he was right! *I HAD been negotiating all my life, and I didn't know it.* How could that be?

My mind began silently racing, and I tuned out the next half hour of whatever was being said. How was it that I, a seasoned 46-year-old professional manager with three engineering degrees and successful by many measures, could only find out now that I always knew how to negotiate? And that I had been silently afraid of this demon for over two decades? What was the reason?

The somewhat simple answer came to me that morning: no one had ever taught me this—not in engineering school, not in my work career—at no time in my life had somebody drawn me aside and told me about this important skill.

> It is very important to get an idea of how things really work outside of school and hearing about this "stuff" would have been helpful to me when I graduated.
>
> —Business manager attending Tau Beta Pi seminar

I then started thinking about my engineering education, and began getting angry. To be sure, I had a wonderful education at three excellent institutions—The Cooper Union, Yale University, and Polytechnic Institute—but now, looking back, it was virtually all in technical education—science and technology—not "negotiating" nor, as I thought about it in the class, many other skills like public speaking, writing, making decisions, setting priorities, dealing with meetings, working with other people, and so on. They didn't, and still don't, teach you this stuff in engineering school: the non-technical "soft" skills so important in the real world of work and life.

> Engineering schools do an inadequate job at producing well-rounded engineers with background courses in English, writing, economics, history, and so on. The trend also appears to be against this, that is, more technical and managerial. The latest push industry-wide toward "continuing professional development" also emphasizes the technical side, that is, they are not counting CEUs in nontechnical courses. Therefore, the burden is going to fall on the employers, who, by and large, are engineers with the same weaknesses."
> —Engineering manager, Port Authority of New York and New Jersey

I began to make a list, right then and there, of all the things I didn't learn in engineering school, and it quickly became about two pages, single-spaced. So that's how *Stuff You Don't Learn in Engineering School* was born.

"So what?" you may be asking. Maybe this personal experience happened only to me and perhaps to no one else in the world, given the infinite variations in our personal lives, education, work experiences, and personalities. However, it would not surprise me to learn that many if not most engineers who have gone through engineering school and the early parts of their careers can relate to this. Today's "real world" is so complicated and fast-changing, and engineers are concerned not just with technical competency, but with so many things involving work and family that their engineering education has not prepared them for.

To be sure, engineering schools provide a magnificent education, certainly very tough and demanding, and they do attempt to teach some of these soft skills in more robust programs. But, let's face it, it's just not like the real world of work, especially with the emphasis in school on teaching more and more technical subjects. It seems that most other college educations with more well-rounded curricula help better

Figure 1.1   Welcome to the "real world," where you're expected to know your stuff and get things done.

prepare graduates for the working world that demands skills of both knowing your subjects *and* dealing effectively with people—your manager, coworkers, clients/customers, elected officials, and the public.

So this book is less about "Why can't engineering schools give engineers a more effective education?" and is devoted instead to helping engineers become acquainted with the most important nontechnical "soft" skills not taught in engineering school that are necessary to be more effective and happier persons. So let's get started!

> Generally, the real world requires a person to be a well-rounded individual. Lacking a certain skill puts [engineers] at a disadvantage.
> —Engineer, major construction company

WELCOME TO THE "REAL WORLD"!

The much-talked-about "real world" is many things to many people, but it certainly is different than engineering school! Even if school is a very difficult time of your life, it is still like a cocoon, in that you are protected from the time when you need to find a job and support yourself and perhaps others. Young engineers have many concerns about the real world; see the Appendices for some fears raised about the uncertainties, expectations, and difficulties of dealing with projects and people.

> They definitely don't teach this [stuff] in school. . . . It's not something I would have heard elsewhere.
>
> —Engineer at Tau Beta Pi seminar

Although nothing I say can completely settle your worries about the future, I happen to be very optimistic that we live in a wonderfully stimulating time for professionals in engineering and technology. There are no assurances, but there are so many positive factors in the economy and society, and, even with all the troubles in our world, I believe that one would be hard-pressed to identify another time when it was better to live.

Given that, it is important to realize that *you have no control over the real world*. Change in our world, and its impact on our lives, is a constant, and the acceleration of change in recent decades has been unnerving to many in our society. The fact that you as an engineer have a lot to do with this by applying advances in science and technology does little to calm your fears about the future. So, if we can't change the real world, we have to realize that *we can control our behavior and actions in dealing with the real world*.

And there is *never enough time* in the real world to do all the things you need or want to do. The days of pulling "all-nighters" to finish papers or cram for exams are over. You can no longer afford the time to make things perfect; indeed, you "perfectionists" out there (I was one!) must understand and accept that you can never achieve perfection in your work. You must accomplish the best job you can within the time allotted.

> What exactly is the real world? If you're talking about whether college has prepared me professionally, technically, and academically for the real world, then I would have to say, not entirely. I don't think I could ever be prepared for it unless I actually experience it first-hand.
>
> —Senior engineering student

THE REAL WORLD IS NOT NEAT

It is not linear, where you can always plug in numbers in an equation to get a "right" answer. Information you need is not handed to you on a plate. It is not like doing Problem 2.4 in the textbook. People aren't standing there at your beck and call to answer your questions on this or that. Technical issues you're dealing with often must be factored in with policy issues and people issues. Information you need is often imprecise, if it is available at all. Things will pop up that affect what you are trying to accomplish, adversely as often as they may represent a positive opportunity. One of many "Murphy's Laws" says, "If it *can* go wrong, it *will* go wrong," and you probably have little or no control over any of this.

> [Thinking about the real world] made me realize that it is not good enough to be technically competent. One needs other skills to succeed.
> —Engineer at ASCE seminar

So what is an engineer to do? For one thing, understand that there *is* life after engineering school. You are not alone; engineers have preceded you in entering the world of work and have, for the most part, adapted and succeeded. But do you feel well prepared to do this? Probably not. Let me, therefore, give you some encouragement.

Think about this question for a moment: Do you someday aspire to become a Chief Engineer or a Chief Executive Officer (CEO)? When I ask this of young engineers at seminars I give, about one-third raise their hands. Some hands shoot up and some rise rather haltingly. The other two-thirds seem a bit uncomfortable. Perhaps they feel embarrased that they have such lofty aspirations or are hesitant to appear so ambitious to the others. However, believe it or not, *each of you is already a CEO.*

YOU ARE THE CEO OF YOU!

You are responsible for your behavior, the decisions you make, your life choices. You and nobody else. Sure, you can and should get advice, but you are the one who is ultimately responsible. That's what a CEO does. So start thinking of what you want to do in life, what actions you will take to make your life better. Don't defer to others to make important decisions in your life. If you do this, you will be well on your way to being more effective and happier in the real world.

GOALS FOR THIS BOOK

What will we try to do in this book? What will we try to accomplish that will be worthwhile to you in your engineering career and in your personal life? I'll boil it down to the following three broad goals:

1. *Acquaint engineers with real-world issues.* Become aware of the most important soft, nontechnical skills that engineers need.
2. *Help build your strengths, but target on your weak areas.* We all tend to do well the things we're interested in, but avoid the things we're not good at or afraid of. Hopefully, this book will give you an understanding of all these skills and ways that you can improve in all areas.
3. *Be more effective and happier in work and life.* I think that this goal derives from the first two: if you are aware of all the soft skills that are important, and you work to improve your abilities in them, you will be more effective and happier in your work and in your personal life. You will be able to accomplish much more, and do things faster, so there will be more time for other things in life.

A few more final introductory thoughts follow. First, I do not subscribe to the old adage that "What you don't know won't hurt you." Rather, I strongly believe that "What you don't know *will* hurt you, or hold you back." You've gone through one of the toughest educations in engineering school, and you need to commit yourself to life-long learning, so don't feel that there are some areas that you don't need to know to get by. Take, for example, public speaking. Many young engineers feel that they don't have to learn how to speak in front of groups. This is flat-out wrong. Any young engineers who feel that they can succeed in their careers without adequate speaking, writing, and people skills have to adjust their thinking. That's what we talk about, and teach, throughout this book. Not just the importance of the skill, but *why* it's difficult for engineers and the *how to do it* part as well.

The nontechnical skills covered in the book are not in any particular order as such. Remember, the real world is not neat and linear. So you can read the chapters in almost any order you prefer—from start to finish, or dip into various skills that you need first, and certainly go back to refresh yourself anytime you need to. It even may be helpful to think of these nontechnical skills as being like a lunch buffet of "stuff" that you need to know. At a buffet you see all kinds of interesting foods, many of which you are familiar with and like or dislike, and some other things that may look like "mystery meat" without the little signs to identify the food. So think of the nontechnical skills we will cover as somewhat analogous to the lunch buffet—you're better at some skills perhaps because you like them already, you avoid some things because you think you don't know how to do them or you dislike doing them, and some other things you won't even try because they look too unappetizing. Well, I'm not your mother who, like mine, urged me to just try things—"How can you say you don't like it if you haven't tried it?" Does that sound familiar? My goal is to give you the incentive to "try things," the important nontechnical skills, and then—to your surprise in many cases, I'm sure—you will like it and get good at it!

As you read this book, please keep in mind and relate the material to the things happening in your work and in your personal life: important decisions you need to make, handling your "to do" list, focusing on important projects, and dealing more effectively with meetings you're going to or leading. One of my personal pet peeves when I went to professional development courses was that they were almost never tailored to my work situation. I left all my work piled up back at my desk, and only learned generic skills—say, about new software or public speaking. Usually, I got a big looseleaf from the course, which I promised myself to read and refer to after the course, but rarely did. I would return from the course with more work added to the pile, phone calls and emails to return, and not too happy about the time spent away.

Make this book help you with things that are on your plate right now and in the future. Use it as a kind of "cribsheet" or "cheatsheet" and refer back to it before, say, you run a meeting. Highlight key tips, dog-ear the pages, *use it*! In my first summer job a long time ago, I was thumbing through an engineering manual that a senior engineer had, and it was well worn. It turned out that it was the second or third copy that he had bought, showing that it got a great deal of use. So I went out and bought this useful book and still have it on my bookshelf. In the same way, I hope this book will stand the test of your time and application.

FIND OUT WHAT WORKS FOR YOU

This is so important. Each of us are individuals with different backgrounds, abilities, and interests. One size does not fit all. So see what works for you that meets *your* needs. I'll offer numerous tips and suggestions along the way, many that I've gotten from asking others how they dealt with certain things. Frankly, I've become a very open "sponge" to learn all sorts of ideas and ways of handling things. For example, one time I asked a terrific engineer how he managed to remember to follow up with me on things, some of which I had forgotten, and he whipped out his "steno" book—the small spiral-bound notebook that secretaries used to use to take shorthand or stenography and transcribe into writing. After that, I began to use steno books to take notes at meetings, note follow-ups, and such; and used this for years until I recently shifted over to using a Personal Digital Assistant (PDA), discussed later in Chapter 6 on "Setting Priorities."

> The development of these [soft] skills must be balanced with technical development. Recent trainees that we have hired have great computer skills, which are critical to developing designs and drawings. The trainee program should [also] include seminars for leadership training, public speaking, presentation, and technical writing.
> —Assistant chief architect, Engineering Dept.,
> Port Authority of New York and New Jersey

Last but certainly not least, *remember that technical competency is the core of your success in engineering.* Although I maintain that learning the nontechnical soft skills is so important to supplement and complement your career, it is crucial to maintain and enhance your technical abilities in the fast-changing world of technology. That is a job in itself, but one that you are probably very well suited to handle; through continuous learning, additional edu-

> **Stuff You Don't Learn in Engineering School**
>
> • Don't be afraid of something. Learn about it!
> • You are responsible for yourself: *You are the CEO of you!*
> • Get to know the most important nontechnical soft skills.
> • Build on your strengths, but target your weak areas.
> • What you don't know *will* hurt you, or hold you back.
> • *Remember: technical competency is the core of your success in engineering!*

cation, membership in a professional society, and keeping up with your field.

Enough said for openers. Let's move on to the stuff you don't learn in engineering school!

SUGGESTED READING

At the end of each chapter is a relatively brief selection of readings on the various themes from each chapter. Note that many of the readings cover broader themes that are covered in other chapters, as the material does not always neatly fit into a single topic or skill area. Readers are urged to find additional resources by searching topics on the Internet, browsing at bookstores, reading professional society publications, and, not to be forgotten, making regular visits to the local public library. My library card is the most valued possession in my wallet.

Adams, Scott. *Dogbert's Management Handbook.* New York: Harper Business, 1996.

Adams, Scott. *The Dilbert Principle.* New York: Harper Business, 1996.

Florman, Samuel C. *The Existential Pleasures of Engineering.* New York: St. Martin's Press, 1976.

Florman, Samuel C. *The Civilized Engineer.* New York: St. Martin's Press, 1987.

CHAPTER 2

WRITING

> Communication comes in many forms in life—both in the workplace and at home. Meetings, memos, speeches, even simple conversations at lunch, all require strong communications skills to convey the proper message. Yet, engineering schools seldom teach such fundamentals.
>
> —Mechanical engineer at ASME seminar

Early in my career, I returned to my office with my manager and he told me to "write up" the meeting we had just attended. "Write what?" I asked. "You know," he said, "just write a memo about what happened. Who said what, what was decided, next steps, and so on." I returned to my desk and stared at a blank piece of paper for a very long time, not knowing where to begin and what to say. Today's version might be staring at the blank screen of my computer, trying to begin writing something. Do I need to say that what I wrote so long ago was pretty awful, not to mention the stress I went through?

This is the first of two chapters about communicating better, by which I mean writing, speaking, and listening. These so-called "communications skills" are so prized in today's work world that engineers can be at a real disadvantage if they are not proficient in these areas. This triumvirate of skills has been called the "Achilles' heel of engineers." And with good reason: we didn't have to do much writing or speaking in engineering school, and we may not be called upon to do much on the job. But that doesn't diminish their importance: without them, you'll never be able to convey the merits of your work, ideas, and aspirations to people, whether they be your manager, your clients, your family, or the public.

Why is it that engineers often have so much trouble writing, whether it's a report, a technical memo, a business letter, even just a note to a colleague? So let's start our discussion about communicat-

Stuff You Don't Learn in Engineering School. By Carl Selinger.
ISBN 0-471-65576-7 © 2004 the Institute of Electrical and Electronics Engineers, Inc.

Communication skills (writing, public speaking, etc.) seem in general to be sorely lacking with young engineers. It seems more and more colleges are focusing on technical education and minimizing the liberal arts portion of the curriculum. While the technical end of this profession gets more and more "complicated" with advances in technology, effective communication becomes more demanding. Higher percentages of time are now spent on communication and report-writing, and often I have seen poorly written reports lacking coherence.

—Engineering manager, DMJM+Harris

ing better with writing because it is often cited as the single, quintessential weakness of engineers that they can't write. Well, perhaps they can.

First, a few anecdotes. I once interviewed people for the position of secretary of our unit. Most of the candidates were very well qualified and some had a college degree which, to me, meant a higher level of ability, but the person I hired was Kay, who had only a high school education, after she told me that she wrote all the letters and reports in her engineering unit. "Really," I asked, "why was that?" "Because," she said, "the engineers there didn't know how to write."

Another time, the director of my department continually bemoaned the "poor writing" ability of the professionals, including many engineers. Over time, it turned out that it wasn't so much that their writing was that bad; it was that they weren't able to clearly express their thoughts to the director, or at least be in conformance with the director's ideas. He was also somewhat of a perfectionist, a trait not uncommon among control-oriented executives. But, to be sure, clear writing can reflect clear thinking, and vice versa.

So, although writing better is an important skill, we come to it with different levels of ability and have to write in so many different situations. Let's start by asking a question: *What things do you write these days?* Your list would probably include some or all of the following: status reports, technical memos, specs, bid documents, business letters to clients or the public, memos, lab reports, meeting minutes, transmittal letters, writing comments on drawings, and, of course, e-mail—a *lot* of e-mail. Writing those kinds of documents may not require the same level of talent as, say, a novel, but it still takes skill. Each uses a different writing mode, with its own rules and formats, and has one or more of the following goals: to inform, to explain, to persuade,

Figure 2.1   Staring at the blank screen. What are you going to write? A memo, technical report, or plain old e-mail?

and/or to entertain. The key to better writing is to accomplish one or more of these goals using the least amount of verbiage.

To begin, *always use clear, simple, direct language.* Don't feel like you must constantly use technical language, jargon, or acronyms. Highly technical language certainly has its place in today's world, but often your audience is not technical—for example, a manager, a sales person, the public, or an elected official. Acronyms can be especially troublesome and confusing, even though the temptation is there to use this technical shorthand over and over.

Be careful that your audience understands the acronym; you will not be there to tell them what it means. One time, a colleague called me because she got an e-mail and didn't know what was meant by the acronym "CRM" in the context of information technology at airports. I didn't know what CRM meant either, and, after we took some creative guesses, I suggested that she reply to the sender and ask what it meant. This made her uncomfortable because she felt that she probably should know what it stood for. (Readers: Do you know what CRM stands for?) The simple way to deal with this is to spell out the acronym the first time you use it. For example, use Customer Relationship Management (CRM) (the answer!)

Figure 2.2   Written comments on design drawings need to be clear, like all types of writing.

initially. Then you can use CRM the rest of the time. Use of too many acronyms and technical jargon can confuse everyone, even your technical audience. Strive for clarity and simplicity in your writing, especially in the initial topic sentences, and then feel free to "drill down" into more technical writing as you go along.

A GREAT WAY TO IMPROVE YOUR WRITING IS TO READ MORE

Yes, read more! If you don't have the time, then *make* the time. Almost any kind of reading is worthwhile—a good newspaper like *The*

> Something good has come out of the computer revolution. The "spell check" eliminates the worst of grammatical, punctuation, and spelling errors. Now if they only knew how to write.
> —Engineering manager, The Port Authority of New York and New Jersey

New York Times, a novel or nonfiction book, magazines like *The New Yorker* and *The Economist,* whatever interests you. The more you read, the more you will see and appreciate good writing. And, over time, it will rub off on you, improving your vocabulary, your grammar, and your story-telling ability. Plus, you'll probably enjoy the relaxation that comes with reading.

Another tried-and-true technique is to *get a peer to critique your writing.* You may feel uncomfortable about asking. I mean, doesn't this mean that you don't know your stuff? You shouldn't feel this way. You need to be confident enough to ask for help. Say you've been laboring over a report, but since you've seen it so many times, you cannot tell if what you've written is clear. A friend or colleague will be able to tell you if they understand what you're trying to say, and maybe point to sections that are unclear and need revision. Whenever I'm stuck on a passage, I imagine my former manager, Bud, reading what I've written, asking me a question about an unclear point, listening to my response, and then saying: "Write *that!*" I also use that technique mentally when I'm stuck on how to start something. I pretend I'm explaining it to Bud. (Thanks, Bud!)

How to Write Better

- Use clear, simple, direct language.
- Minimize use of acronyms, and be sure to explain them the first time.
- Read more, to see good writing and improve vocabulary.
- Ask peers to review your writing.
- Mentally pretend you're talking to someone, and write that!

Your writing will improve if you try some of the above suggestions. If you need more help in developing your writing skills, take a course, read a book, or talk to your manager. Like all the nontechnical skills we will cover, this is not an exhaustive "how to" book; it is meant to acquaint you with the need for good writing skills, give

---

I need to develop my writing and speaking, to gain the ability to speak what I believe freely, not worrying about the negative impression it could possibly give people.
                                        —Engineering student, Webb Institute

you a sense of why engineers might have particular difficulty with writing, make some suggestions to help you write better immediately, and tell you where you can go to find more help.

The bottom line here: engineers must be able to write effectively. Period.

SUGGESTED READING

Andersen, Richard. *Powerful Writing Skills.* New York: Barnes & Noble, 1994.

Bailey, Edward P. *The Plain English Approach to Business Writing.* New York: Oxford University Press, 1990.

Beer, David (Ed.). *Writing and Speaking in the Technology Professions: A Practical Guide,* 2nd ed. Hoboken, NJ: Wiley/IEEE Press, 2003.

Beer, David. *A Guide to Writing as an Engineer,* 2nd ed. Hoboken, NJ: Wiley, 2004.

Blicq, Ron S., and Moretto, Lisa A. *Writing Reports to Get Results: Quick, Effective Results Using the Pyramid Method,* 3rd ed. Piscataway, NJ: IEEE Press, 2001.

Hirsch, Herbert L. *Essential Communications Strategies: For Scientists, Engineers, and Technology Professionals,* 2nd ed. Hoboken, NJ: Wiley/IEEE Press, 2003.

Lauchman, Richard. *Plain Style: Techniques for Simple, Concise, Emphatic Business Writing.* New York: AMACOM, 1993.

Quattrini, Joseph A. *Clear and Simple Technical Writing.* Englewood Cliffs, NJ: Prentice-Hall, 1986.

Safire, William, and Safir, Leonard. *Good Advice on Writing: Great Quotations from Writers Past and Present on How to Write Well.* New York: Simon & Schuster, 1992.

Sageev, Pneena P. *Helping Researchers Write . . . So Managers Can Understand.* Columbus, OH: Battelle Press, 1994.

Scott, Dewitt. *Secrets of Successful Writing: Inside Tips From a Writing Expert.* San Francisco: Reference Software International, 1989.

CHAPTER 3

SPEAKING AND LISTENING

PART I—SPEAKING

> I need to develop my speaking skills, speaking up in meeting
> situations, but, generally, I am not assertive enough.
> —Engineer at Rice University seminar

I CAN STILL see them falling asleep. It was over 30 years ago,
and they still nod off in my mind's eye, one by one. I remember
being devastated. I was trying to tell them how the surveys we were
doing would measure the effects of staggered work hours on
transportation systems. But the group of business people from the
Broadway Association weren't very interested and they were telling
me this by closing their eyes, and it was only ten o'clock in the
morning. Even my boss had his eyes closed! I now realize I had
committed a cardinal sin in public speaking—not knowing what my
audience was interested in hearing.

For many people—not just for engineers—there is nothing
more scary than public speaking. Surveys show, unbelievably, that
people are more afraid to speak in front of a group than they are of
dying! I personally think that this is a needless waste of fear and
worrying, since public speaking is something that everyone can do
well if they would just learn the skill.

> Writing and public speaking are the two most important soft skills that
> engineers need to succeed in the field today. There are many engineers
> who can perform the technical work, but being able to convey that tech-
> nical work to the client and/or public is what allows engineers and the
> company they work for to prosper.
> —Engineering manager, DMJM+Harris

Stuff You Don't Learn in Engineering School. By Carl Selinger.
ISBN 0-471-65576-7 © 2004 the Institute of Electrical and Electronics Engineers, Inc.

Figure 3.1 Briefing a meeting about your project is critical to informing or persuading people about your ideas.

I must tell you that when I got out of engineering school, I could have been the "poster child" as the worst public speaker in the world. No one was as bad as I was. I was petrified of speaking, and when I had to, I spoke much too fast, mumbled and slurred my words, and kept on going and going, not knowing when or how to stop talking. Pity my poor audiences. Now, after learning how to speak, as I'll share with you shortly, I enjoy speaking in public and it is a critical skill in my work.

Simply put, you, as a young engineer, *must* be able to speak in front of a group of people. Now I'm not talking about your delivering the State of the Union Address, or testifying before the Congress, or giving a major speech in front of thousands of people, at least not at the start of your career! I'm talking about the many and varied times when you will have opportunities to speak. These include voicing your ideas at a staff meeting, briefing your peers after a business trip, pitching a proposal to clients, asking a question in a meeting, presenting a paper at a technical conference, and discussing a project to the general public or an elected official.

Try to convince young engineers that it's OK to ask questions.
—Chemical engineer at AIChE seminar

And even if you are sure that you are the most tongue-tied, introverted engineer around, I am here to tell you that you *can* learn to speak better. If you can learn those really hard subjects in engineering school (or at least get passing grades!), you can learn how to speak in public. Think for a moment about the skill of driving a car. We aren't born knowing how to do this highly dangerous and complicated task. We *learn* how to do it, and then over time get better at it by practicing. Ultimately, some of us are better drivers than others, and some of us enjoy driving more than others do. But it's a skill that can be learned. The same goes for public speaking. But this doesn't happen by osmosis, you do have to learn some basics about speaking.

A first rule of speaking is that you must *know your subject*. Usually, this is not a problem because that's why you've been asked to speak. However, if you're asked to address a topic that you don't feel comfortable speaking about, you should politely decline the offer. If that's not an option, then do as much research as you can beforehand, until you feel reasonably comfortable with the topic.

Next, it's crucial to find out *who will be your audience and what they expect to hear.* The people who invited you to speak can probably furnish that information. Your presentation should be prepared with this in mind.

Figure 3.2 Preparing a PowerPoint presentation is so easy. It collects your thoughts and converts them to slides and handouts in an instant!

Preparing a presentation today and using audio-visuals has never been easier. PowerPoint has become the speaker's audio-visual aid of choice, so you should learn how to use it. If you find it's not for you, investigate other visual aids for your talks. Also, there will be many times at which you just need to have notes to refer to and not a formal presentation. If you will use any equipment for your presentation, make sure you know how to use it and arrive earlier to set up any equipment before you take the stage. The sight of you fumbling with a PC projector and their laptop connection will not reassure the audience.

Now, this is really important: *Never, ever read your presentation*. At the least, you will bore your audience. Know your stuff, practice sufficiently, use your PowerPoint bullets as your on-screen notes, and you won't need a prepared script. Be conversational, look at people in your audience, connect with them. The minute you start to read something, you turn them off. So don't use reading as a crutch—you do not need it.

It almost should go without saying that you need to speak in such a way that you can be understood. This means that you should speak loudly enough to be heard, whether with a microphone or just your voice. If you're not sure you can be heard, ask the last row of the audience to raise their hands if they can hear you. Speak slowly enough so you can enunciate words clearly; many people have the habit of speaking too fast, especially if they're a bit nervous. So, slow down! And lastly, if you have an accent that makes it difficult for people to understand you in normal conversation, then you need to make a special attempt to speak loudly and slowly, and even ask the audience if you are speaking clearly. (If you have a serious problem with people understanding you, I suggest you look into getting a speech coach.)

Two other personal traits can be essential for an effective talk. First, you need to be *sincere* about what you are talking about so that the audience will believe in what you're saying. Or, conversely, why should they believe what you're saying if you don't seem to? Further, try to be *enthusiastic* about your subject; again, why should the audience care if you don't?

It is crucial to *keep to the amount of time* you have been allotted for your talk, if only to allow adequate time for a question-and-answer period. There is nothing worse than watching the audience start to squirm as you continue to speak way beyond your allowed time. You'll probably recall some professors who kept droning on even after the class practically gets up and walks out. Even a famed public speaker like President Bill Clinton committed an egregious speaking error when he was Governor of Arkansas. His seconding speech, in support of Governor Michael Dukakis's nomination for President, kept going on and on (despite people off-camera frantically waving their hands at him to stop!) so that the news of Dukakis's nomination missed the critical 11 PM evening news on the east coast. A more common abuse of allotted time that you may run into is a panel discussion in which each presenter is given a certain amount of time, say 20 minutes. If you're the last of four speakers, and the first three panelists each go over by 5 minutes (because the moderator is not skilled in controlling the time), then you're practically out of time before you start (and not a happy camper).

How do you keep track of time? The easiest way is to use a clock, if available, and, if not, use your watch. Take it off and put it on the lectern where you can see it easily. Focus on when you need to *finish* speaking; that is, remember the time you have to stop and then work back from that so you can finish in your allotted time. Also, you can even ask someone in the audience to signal you when you're approaching your time limit. Keeping to time may be almost as important a speaking skill as knowing what you're actually talking about!

> I need to develop my speaking skills in formal presentations. I'm comfortable speaking in front of a large group of people in an informal setting, but when I have to formally present something, I get pretty nervous. No matter how much I prepare, I don't feel like I prepared enough.
>
> —Engineer at Tau Beta Pi seminar

After your talk—and if you've kept to time!—you should have time for questions, and be ready to answer them effectively. To some people, this can be the scariest time because you don't know what to expect. However, it can be the best time for the audience, now that they have a chance to engage with you. Many people have difficulty answering questions; in the worst case, they go on and on and don't even answer the question. A good way to answer questions is to use the "PREP" method: make your Point first, then give the Reason, followed by an Example or two, and then repeat your Point. I'll illustrate the PREP method by answering a question to cover my next point.

"What happens when you don't know the answer to a question?" Don't be afraid to say "I don't know" if that's the case (Point). Everybody doesn't expect you to know everything (Reason). You can give an example or two of what the questioner may find useful (Examples). And then you can say that you're sorry you weren't able to answer their question but that you'd be happy to get back to them if they give you their business card (Repeat the Point).

Let's discuss how to deal with nervousness. First, it is very natural for most people to be nervous before speaking to a group, even experienced speakers. It is good to have an "edge" in that it forces you to be sharp and awake. But there is an easy way I've learned among the many tips for controlling nervousness; it's called "diaphragm breathing." Basically, it's a way of taking several deep breaths with your stomach going out—*not* with your chest expanding—a few minutes before you're going to speak. Your lungs will fill with more oxygen, your voice will get lower, and you will feel more comfortable. I do this all the time—while sitting on a dais and being introduced, or even when going up in an elevator before an important meeting. The beauty of this is that no one notices you're doing it, and that it works! Try it.

Above all, *practice, practice, practice!* Practice your public speaking. Don't wait for an invitation; seize opportunities to speak. Some examples: offer to introduce speakers at meetings, give briefings on your projects at staff meetings, join your organization's "speakers bureau." The more you practice your skills, the better you will become, just like driving. For those who feel they need more help, consider joining the nonprofit group Toastmasters International (http://www.toastmaster.org), which has chapters throughout the world where members meet regularly and practice speaking.

The Essentials of Public Speaking

- You CAN do it! Just learn how and practice!
- Know your audience—what do they want/expect to hear?
- Keep to time.
- Be sincere and enthusiastic.
- Practice, practice, practice!
- Never EVER read!
- Speak slowly and clearly, and loud enough for all to hear.
- Join "Toastmasters."

Public speaking is a crucial skill you need to learn. And once you learn, and the more you speak, the more comfortable you'll be, and the better you'll get. And I bet you even learn to enjoy it!

PART II—LISTENING

"You're not listening to me!" Do people often say this to you? It happens to me quite a lot. For many of us, really listening to someone can be the most difficult aspect of communicating with another person. It is not as simple as it sounds, and that may be why we are not too good at listening effectively. You can certainly *hear* the other person talking, but are you actually *listening* to what they are saying? The good news is that listening is a skill that can be learned, but you do have to work at it.

First, you need to realize that listening requires *paying attention*—concentrating on what the other person is saying. We think that just by hearing the other person speak, our brain records everything that's said. Not true. In addition, while we're listening, we're also thinking about that project that's overdue and what we're going to have for lunch that day. Our brains are constantly multitasking while the other person is speaking, and, therefore, we have to discipline ourselves to concentrate on what the other person is saying. That is easier said than done.

Listening requires not just paying attention to the facts being conveyed by the speaker, but to their feelings, ideas, and overall intent. This may be expressed by their words as well as through their mannerisms—their facial expression, body language, and the tone of their voice. Although we may get caught up in trying to remember the details—the numbers and the facts—it may be more important to understand their overall meaning. You need to tell if they are they happy, or they are angry with something—perhaps you—or if they are trying to convince you to do something.

There are some good ways to improve your listening skills. First, try to *reduce or eliminate any distractions* when you are speaking with someone so that you can concentrate on their words. If you're conversing with someone in person, try to *keep eye contact with the speaker* so they see that you're paying attention—a moderate amount of eye contact, don't stare at someone. Why is eye contact so important? If you're looking away—at something else, per-

Figure 3.3 Make good eye contact when speaking with someone to show you're listening. Looking away from the other person might signal to them that you're not paying attention or lack confidence.

haps the scenery or other people walking by—the other person may get very frustrated or angry with you since it *appears* that you're not listening to them. If you look away and avoid eye contact completely, perhaps by looking down, it could be a sign to them that you lack personal confidence.

While listening, make sure you reassure the other person that you are understanding what they are saying and respond appropriately, for example, nod your head in agreement, say "I see your point," and perhaps take notes. Be careful not to just sit passively, as the person may become frustrated or angry that you don't seem to be responding.

If you don't understand what the person has said, don't hesitate to say so. ("I'm not clear on that point. Would you explain it again?") This is especially important since the other person will expect you to have understood what they just said unless you open your mouth and ask. So don't feel embarrassed in asking; you may learn something else as the person explains the point again.

Try not to interrupt the other person, or complete their sentences; this can be very annoying and frustrating to the other person. I am often guilty of this bad habit; I just can't wait for people to finish what they are saying so that I tell them what I want to say, which impairs my ability to listen to what they're saying.

Listening to someone who is asking you a question can sometimes be difficult. I always find myself trying to frame my answer before the questioner has even finished. If this is a problem for you, try to let the person finish their whole question since you can always pause before responding. ("That's an interesting question. Let me think a moment before I answer.") Another way to give yourself more time to answer is to reiterate or reframe their question. ("Let me see if I understand what you're asking. You want to know. . . .")

Keys to Listening

- Listening is not just "hearing" the other person.
- Avoid distractions and make eye contact.
- Concentrate on the person's ideas and thoughts, not just facts.
- Reassure the person that you're paying attention; ask questions.
- Confirm that you've understood—"I think you're saying that. . . ."

Not all listening is done face-to-face, of course. You also need to listen while speaking with someone on the telephone. In fact, you can be more easily distracted on the phone since the person is not looking at you (at least not until we all get "picturephones"!). They may even sense that you are not paying attention particularly when you become too quiet during a phone call ("Are you still there?"). The keys to effectively listen on the phone remain the same: minimize distractions, pay close attention, and respect the speaker.

Figure 3.4 Concentrating and avoiding distractions while listening on the phone can be difficult.

I admit, though, I still find listening a tough skill to master. While another person is speaking, my brain is often busy planning what to say next, or I'm just distracted by other thoughts. Becoming a good listener may mean breaking some lifelong habits in how we converse.

So, as we finish our discussion of "communications skills," can you be a competent engineer without being able to write, speak, or listen effectively? Sure, but you will not be nearly as successful as one with good skills. Communicating with people is one of the most basic and important of the "soft" skills, and, with practice and patience, you too can become an effective writer, an interesting speaker, and a good listener.

SUGGESTED READING

Speaking
Beer, David (Ed.). *Writing and Speaking in the Technology Professions: A Practical Guide,* 2nd ed. Hoboken, NJ: Wiley/IEEE Press, 2003.

Beveridge, Albert J. *The Art of Public Speaking.* Los Angeles: Nash Publishing, 1974.

Clinton, Keith. *Facing a Crowd: How to Foil Your Fear of Public Speaking.* Bend, OR: Drake, 2002.

Davidson, Jeffrey P. *The Complete Guide to Public Speaking.* Hoboken, NJ: Wiley, 2003.

Janow, Ellie W. *How to Speak English Without an Accent: A Simple Speech Improvement Course that Gives You the Cutting Edge* (Sound recording). Brooklyn, NY: Sky Parkway Publications, 1990.

Lucas, Stephen E. *The Art of Public Speaking,* 3rd ed. New York: Random House, 1989.

Mandel, Steve. *Effective Presentation Skills.* Menlo Park, CA: Crisp Publications, 1993.

Qubein, Nido R. *How to Be a Great Communicator: In Person, on Paper, and on the Podium.* New York: Wiley, 1997.

Quick, John. *A Short Book on the Subject of Speaking.* New York: Washington Square Press, 1978.

Reimold, Peter, and Reimold, Cheryl. *The Short Road to Great Presentations: How to Reach Any Audience Through Focused Preparation, Inspired Delivery, and Smart Use of Technology.* Hoboken, NJ: Wiley/IEEE Press, 2003.

Westerfield, Jude. *I Have to Give a Presentation, Now What?!* New York: Silver Lining Books, 2002.

Listening

Barker, Larry Lee. *Listen Up: How to Improve Relationships, Reduce Stress, and Be More Productive by Using the Power of Listening.* New York: St. Martin's Press, 2000.

Koile, Earl. *Listening as a Way of Becoming.* Waco, Texas: Regency Books, 1977.

Morris, Desmond. *Bodytalk: The Meaning of Human Gestures.* New York: Crown Publishers, Inc., 1994.

CHAPTER 4

MAKING DECISIONS

I feel better now if I have to make any decision because there are no right or wrong decisions. All I need is to get enough information to help me in making the decision.

—Chemical engineer at AIChE seminar

YOUR MANAGER ASKS you the question: "What do you think we should do?" Are you confident that you can make a sound decision, or are you petrified that you don't know how and where to begin? Your manager could be asking you about the next steps in a big project or about where to go for lunch, but the skill required is the same: decision making.

The skill of decision making is arguably the most important "soft" skill that engineers need to learn. Yes, learn, but it's never "taught" anywhere. Plus we're dealing with a hidden barrier—the *fear* of making decisions. More on that later.

> . . . [Y]ou don't realize how many decisions you need to make every day until you really think about it.
> —Engineer at DMJM+Harris

So, what important decisions do you need to make now, for work or your personal life? Take a moment now to write them down.

Important work decision: _____

Important personal decision: _____

Now, before we go on, let's do a short exercise that I'd like you to *really* think about: Your manager asks you to choose a restaurant to which he/she can take several out-of-town visitors for a business dinner. Assume they are a team of engineers from a company you're working with, two men and a woman, and only one of them has been to your area before. Pick a restaurant you would recommend and one alternative in case that one is unavailable. Make any reasonable assumptions. Now take no more than five minutes to think about this, and write down your selections.

Recommended restaurant: _____

Alternate restaurant: _____

What did you just do? If nothing else, you've picked two restaurants in your area where you might take (business) visitors in the future. But what you really did was *make a decision,* in fact, two decisions, in about five minutes if you followed the instructions (but, hey, no one was looking!). Perhaps you're thinking that this was no big deal, but in fact you unconsciously went through a four-step process to make a decision. If you understand those four steps, you now have a simple template to make all decisions, big and small.

A PRACTICAL FOUR-STEP APPROACH TO DECISION MAKING

Step 1—*What has to be decided?*
The first step is to realize that a decision is needed. Here, it was straightforward: you needed to recommend a restaurant for a business dinner in your area, as well as an alternative restaurant, in no more than five minutes. So there were two decisions you needed to make.

Step 2—*What are the options . . . alternatives . . . choices?*
This is the key step to decision making—identifying your *options*! Without options (or alternatives or choices), you really have no decision at hand! Many times, these choices are readily apparent. I'll bet that several restaurants came quickly to mind that you then considered. Thinking of a half-dozen restaurants would have been a fine scope for this decision; that is, it wouldn't have been appropriate to go through your Yellow Pages and consider dozens of restau-

rants. By the way, it's often very difficult for engineers to limit their alternatives to a reasonable and logical number, and they end up with too many on their plate to investigate. Further, the so-called "Do nothing" alternative is also a valid option many times, though I prefer to give it a more positive name and call it "Continue current approach."

Step 3—*What information do you need to make a decision? What criteria will be used?*
Not much information was given to you, just a general situation in which you wanted a quality restaurant for a business dinner, even one that was somewhat pricey. In fact, maybe you had too much information that distracted you from what you needed to make the decision. Part of the difficulty of making decisions in our real world is that we get too much information—an "information overload"— that makes it more difficult to identify what we really need to make a decision. Sometimes, the criteria you would use are straightforward: lowest cost, best performance, and so on. Other times, it may include qualitative or subjective criteria such as aesthetics. The "real world" and decision making are not that neat and linear, but there is closure. How do you know that you have *enough* information? You need to gather only as much information as you need to make the decision, based on your judgment, not all the information that you could possibly gather if you had unlimited time and resources. For example, you wouldn't necessarily need to get the menus of all the restaurants you were considering.

Step 4—*Decide! Make a decision! Be decisive!*
OK, the key last step: Decide! Pick one! Actually, you needed to pick two! Was that really that hard? I don't think so. But you needed to make a decision in the time allotted. And, hopefully, you did. This is the most important point: that you need to feel comfortable with actually deciding and then moving on to implement the decision.

> I need to develop my decision-making skills. Engineers are *trained* to *optimize* decisions. *Gathering data* can become the trap to *making a decision.* Not because of fear, but because we need all the data (or so we think). We need help to break that cycle.
> —Engineer at major Midwestern engineering company

Figure 4.1   Thinking through your options is not easy, but is key to making a decision.

At this point I will confess that I left out one factor that, perhaps, is the single most important reason that we're uncomfortable making decisions—*fear.*

I was speaking to engineering students at the University of Connecticut and had just finished a discussion on decision making. Down in the first row, right in front, I noticed a young woman who seemed like she was in pain, her face in a big grimace. I was concerned and asked her if she was OK, and she blurted out: "I'm *afraid* to make decisions!"

> I'm *afraid* to make decisions!
> —Engineering student, University of Connecticut

Yes, the fear of making decisions often holds us back! Let's think about this a moment, because it might be the single factor that inhibits you from being more decisive, even after having learned the step-by-step process.

It boils down to the fear of the humiliation of making a "bad" decision. American culture is replete with people being blamed and

I recognized several flaws in my decision-making processes that needed to be corrected. (Particularly, I realized that putting off a decision involves making a decision in and of itself.)
—Civil Engineer at The Port Authority of New York and New Jersey

ridiculed for making bad decisions on restaurants, movies, and projects: "Who *picked* this place? . . ." "What *were* you thinking!? . . ." "That was an *awful* decision!" So where is the incentive to be more decisive? Good question.

How to Make a Decision—A Four-Step Process
1. What has to be decided?
2. What are the options, choices, and alternatives?
3. What information do you need; what criteria will be used to evaluate?
4. Make a decision! Be decisive! Don't be afraid!

Let me tell a story about having to deal with a direct challenge by an executive to a decision, and how paraphrasing the four-step approach in explaining why we made the decision was so helpful. I was involved with a group that organized our office holiday party every year, and one year we picked a small restaurant that was pretty funky (that is, a bit dark and dirty). When I finally sat down with my meal after an hour, an executive sitting next to me challenged me in a harsh tone: "Carl, who *picked* this place?" (Translation: "Carl, How could you have picked this dump?") While others might have quivered in fear, or gotten defensive, I simply responded (and note how it tracks the four-step process): "Well, we had to make a decision in November, and we had identified several places—where we had the party last year, or another restaurant, like this one, where we had a lunch for Henry. Everyone loved the food here, and we were able to have the whole restaurant to ourselves, so we decided to have the party here. If you have other suggestions for next year's party, please give them to me so we can consider them." The executive didn't have anything to say except, "Oh, I see." End of story.

The most helpful part of making decisions is not being *afraid*! I fear the consequences but I can't look back.
—College engineering senior at SWE seminar

Now a few words about whether there are, in fact, "good" decisions or "bad" decisions. I know this is in the common vernacular, but I don't find it helpful, and here's why. One makes a decision based on the best information and expectations at the time of the decision, but things almost never work out exactly the way one thinks or hopes. However, nothing says that you can't revisit a decision if and when things are not going the way you anticipated. Decisions are not set in stone most of the time. That's not to say that the more important the decision, the more time should be spent in the process of making the decision and then monitoring events closely. I will, in closing, agree that insufficient information or making "snap decisions" or "shooting from the hip" probably can result in decisions that don't work out well.

"Being decisive" is a personal quality that is prized by employers, but rarely if ever will you see it mentioned in a job qualification, on a resume, or in an interview. So become more decisive, get comfortable with making decisions, and be sure to mention this skill at appropriate times, in the same way you'd describe your excellent writing and speaking skills. Becoming adept at making decisions will give you more time to get on with the as-important job of *implementing* the decision, not worrying endlessly or being paralyzed and not deciding. Often, the risk of indecision and inaction is greater than taking action—any action.

> I'm glad to learn about how to make decisions. I'm glad I'm not the only one who has problems making a choice.
> —Engineering student at Tau Beta Pi seminar

But, as a student at one of my seminars once fed back: "Easier said than done!" Except now you know how *easy* it is to make decisions, because someone just taught you. Now, practice!

DECISION-MAKING EXERCISE

This exercise is designed to force small groups to make decisions quickly, in situations in which they could quickly understand what needs to be done, pool their knowledge of information, and seek to make decisions quickly: groups were given no more than 10–15 minutes. The memo format was made to look like this was a "real" situation, using actual people's names, so participants did not realize that this was a decision-making exercise.

Following each team's report on their recommendations, one can use the material in Chapter 4 to reinforce the four-step process to make decisions that they unconsciously used: Step 1—Identifying the decision to be made, Step 2—Determining choices, Step 3—Evaluating the information on each choice, and Step 4—Deciding. The subject that would be decided on depends on the group; this exercise was to choose a restaurant because the group is from the local area and familiar with restaurants in the area. Other subjects that have been used include recommending local sights for visitors to see or what type of food should next year's conference offer.

Memorandum

TO: Carl Selinger
FROM: Gregory S. [actually the VP of the company]
DATE: August 15, 2003
SUBJECT: *CLIENT VISIT TO NY OFFICE: RECOMMEND*
              RESTAURANT
COPY TO: John D. [president of the company]

Would you please have attendees at the "Stuff you Don't Learn in Engineering School" seminar recommend where to have a business lunch on Thursday, September 4th in Midtown Manhattan near our Third Avenue offices, for a visiting team from a former client, a public authority from Cleveland. They will be holding meetings with several engineering firms in NYC to prepare for an airport redevelopment RFQ.

We want to make a good impression on the three out-of-towners on their team: a male middle-aged Chief Engineer who graduated from Florida State; a young Asian woman who is CFO from California State—Chico and who has never been to the New York Metro area; and an African-American airport engineer, who recently retired from Pratt & Whitney in East Hartford, and who has visited New York City frequently.

Please *recommend a restaurant for a memorable lunch* as well as *one alternative* if the recommended restaurant is not available. I will need your recommendation by the day after the seminar. Thanks.

SUGGESTED READING

Allison, Graham and Zelikow, Philip. *Essence of Decision: Explaining the Cuban Missile Crisis,* 2nd ed. New York: Longman, 1999.

Freeman, Arthur M. *The Ten Dumbest Mistakes Smart People Make and How to Avoid Them: Simple and Sure Techniques for Gaining Greater Control Over Your Life.* New York: Harper Collins, 1992.

Okulicz, Karen. *Decide!: How to Make Any Decision.* Belmar, New Jersey: K-Slaw, 2002.

Sloan, Todd Stratton. *Deciding: Self-Deception in Life's Choices.* New York: Methuen, 1987.

CHAPTER 5

GETTING FEEDBACK

What concerns me most about the "real world" is the lack of feedback from most all of the people I work with. In school, I received grades. Here, it's nothing, unless I make a mistake, it seems.

—Engineer at a major Midwestern engineering company

WE'VE ALL PROBABLY had this experience: we hand in a written paper—a term paper, a memo, a draft business letter—and it comes back to us covered in written comments, sometimes in red ink. And we feel really lousy, and perhaps really stressed, and maybe angry as well. Our confidence may be shaken by this other person's evaluation of our work, whether the points are valid or not. We all say that we want constructive criticism to improve ourselves, but when it comes it can be difficult to accept. In engineering school, we have to accept the feedback—the grades are part of the process that leads to a degree. But things change when you get out in the real world.

Note this well: You will never (or hardly ever) get objective feedback in the real world. Sorry, but there are no grades in real life. While you occasionally do get tested on some things, say licensing exams, the days are over for stressing about grades and your almighty GPA (Grade Point Average, if you somehow forgot). This can be a difficult adjustment for engineers who, like all of us, want to know how they're doing. Getting objective feedback can be very difficult or nearly impossible. Yet there are ways to find out what's going well and what needs improving, but you do have to work at it a bit. In other words, you need to try to raise your real-world GPA— of Greater Personal Awareness!

First, let's talk about what is supposed to be the way professionals get feedback these days: something called "*performance ap-*

Stuff You Don't Learn in Engineering School. By Carl Selinger.   **37**
ISBN 0-471-65576-7 © 2004 the Institute of Electrical and Electronics Engineers, Inc.

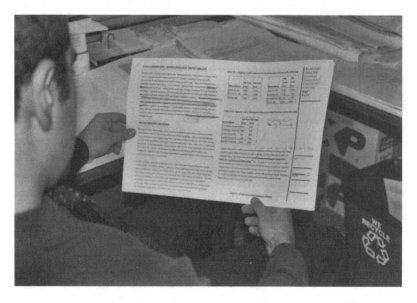

Figure 5.1   A common way of getting feedback is from comments on written materials, which should explain clearly what needs to be changed.

praisals." No doubt, you receive such appraisals from your manager or organization periodically, and at least annually, and they are often tied to your pay and promotion opportunities. These appraisals range from informal exercises to, more commonly, a multilayered series of ratings on your expected goals and performance. On the face of it, these appraisals should be excellent feedback, and, probably in some places in the universe, they are. Not in most places, however. A short anecdote may illustrate this; hopefully, it will not offend any Human Resources professionals reading this.

At a seminar I gave in Hartford with a score of mechanical engineers, I asked for a show of hands by those who felt that their organization's performance appraisal system was helpful. I expected a few people to raise their hands and then share with everyone how this was being done. To my surprise, my question got the biggest laugh that I can ever recall at my seminars. One person thought it was so funny that he almost fell off his chair in hysterics, prompting further laughter by the group. So either my seminars aren't too funny (perhaps) or "helpful performance appraisals" is an oxymoron. (If you don't know what "oxymoron" means, take a moment to look it up—it will be a very useful addition to your vocabulary!)

With that disquieting anecdote, how *can* you get the best feedback out of your organization's performance appraisals? After all, there's a lot of time and effort spent on these, and they are supposed to evaluate your performance against specific goals and accomplishments, as well as rate your strong points and areas needing development—read "weak areas". The key in evaluating performance is to be able to articulate clear work goals and specific accomplishments that are expected, and write them down. (By the way, this is always valuable for you to do to evaluate yourself, even if it's not part of an appraisal.) You need to "be on the same page" with your manager and any other important person you must report to, so minimize any fuzziness or vagueness on what's to be done by you and by when. You must to be able to describe *how* you will know that you've accomplished the goals you set.

Here is an example: One year, my manager and I agreed on my "performance plan" that I would complete a solicitation for proposals to introduce Internet kiosks at major airports, make the business deal, and then implement the kiosks, all by a certain date. This was a clear set of goals and, even though there were the inevitable delays, my manager was satisfied with my performance when it came time to do the appraisal. Anything less clear would present problems in deciding how I did.

At the same time, this manager and I were uncomfortable when it came to his rating my performance, and determining where I needed to develop my skills. This is, to be sure, a delicate and uncomfortable exercise, especially when there is a performance problem. In my example, I took the initiative a bit and asked my manager to identify the one thing he thought I needed to improve during the following year. He responded quickly that I needed to improve on how to negotiate a business deal and suggested I take a professional development course, which I did soon after. This approach was in the spirit of the old adage of "managing your boss," which has many useful applications!

What we don't know can hurt us: so it is important to request feedback from others. What we can't see in ourselves, they may be able to see. Once revealed to us, we can do something about it.
—Engineering manager, TransCore

To be sure, this brief slice of "performance appraisals" cannot begin to cover the breadth of experiences you may have already faced, or will face, in your career. So let's move on to some other practical ways of answering the always-burning question: "How am I doing?"

TALKING TO PEOPLE

Talking to people is a good approach to get feedback. You can learn a lot about how you're doing from certain people in your life, though not everyone. Let's start with coworkers or peers, who have the advantage of seeing you in your work situation and also knowing some of the people, places, and things that you're dealing with, which no amount of explanation can adequately describe to parents and friends. This is very difficult for well-meaning advice givers, since their experiences in life, however valuable, may just not be helpful or correct in your situation.

A good way to get feedback from coworkers, as well as your manager, is to ask questions like:

- "How do you think the project is going?"
- "What would you do in this situation?
- "If you were me, what would you do differently?"
- "What's going well?"
- "Where do you think I could improve?"

Getting usable feedback can be a low-key, nuanced skill. If you try to ask a more direct question like "Am I doing OK?" that practically begs a "Yes" answer, you probably won't get an honest or helpful answer. (When your server at a restaurant asks whether everything was OK, how often do you *really* tell them what you think?) Understand that your coworkers may not be frank or honest with you, or don't want to say something unpleasant, and you may not agree with their opinions and get emotional or defensive. This is a tricky business, so be subtle and probe.

Feedback is motivating, but too infrequent.
—Engineer at Cooper Union seminar

Another approach to getting feedback, especially useful after a presentation you've given, is to ask another attendee whom you trust to check around afterwards for reactions or "buzz." This may include soliciting specific comments from people to feed back to you ("How did you think Carl did?" "Was his presentation clear?"). I once moderated a teleconference at the University of Maryland, and my cousin, a student there, attended and sat in the studio audience. I asked her to listen for any reactions to my moderating so I could get some feedback and make improvements.

Although your parents, spouse, or friends would seem to be good sources of feedback, I believe that this is generally not the case. In fact, their views on your work situation are probably close to useless. Really. Of course, they may be very intelligent and well-meaning. The reason they probably would not be helpful in assessing your work situation is that they would tend to be biased to be overly supportive, which is fine for soothing your ego. You might feel even worse because you know that they will always be supportive, and so lose some credibility to help you in dealing with work situations. If they were critical, you might retort with "You don't understand." And you'd be right; they probably don't—and can't—understand the complexity and nuances of your work issues.

> I ask for feedback but I often cast it off as "He doesn't know anything" and then don't utilize it for my betterment.
>
> —Engineer at DMJM+Harris

Some other ways to get good feedback are ingrained in the normal flow of work. You always need to have your "radar" on for reassurances that things are going well and hints that something is amiss. Try not to be too paranoid with constructive criticism—it is normal to be criticized from time to time, such as with corrections on written work. Just ask for explanations if the comments are not clear. Keep your cool at all times. But remember that the days of a professor giving you marks on your term papers, designs, and exams are over—not that *those* comments were always helpful!

Now that you've gotten some specific feedback, what are you going to do about it? The ball is now in your court. Since someone is suggesting that you do something *differently* than you thought was correct, *you need to consider what the person has suggested.* Although there certainly will be times when you feel that you can

Figure 5.2 Try not to get stressed or defensive by comments on your work. Treat them like constructive criticism and consider each suggestion.

ignore the suggestion, at your own risk, I believe that it is very important to get into a habit of seriously considering *all* feedback and committing yourself to routinely making some changes. For example, I religiously review the evaluations from my seminars and college courses, think about each and every suggestion, and then consciously decide to try at least a few during the next seminar/course.

Getting Feedback in the Real World
- You will never (or hardly ever) get objective feedback in the real world.
- There are no grades in life—sorry!
- Solicit feedback from friends and peers.

So if we really want to know how we're doing, then we need to get feedback, consider all suggestions, and make some changes. And that is the bottom-line reason for exerting the extra effort that is often required to get good, objective feedback in our complex real world—the need to take positive, constructive actions to address our specific weaknesses or deficiencies. Although it's a lot easier for us to work on things we're already good at—in fact we *enjoy* doing those things—it is much harder to discipline ourselves to significantly improve our weaker areas. But the rewards of doing that routinely can enhance your effectiveness and happiness in your work and personal lives.

So, increase your "GPA"—your "Greater Personal Awareness"!

SUGGESTED READING

Bossidy, Larry. *Execution: The Discipline of Getting Things Done.* New York: Crown Business, 2002.

Lazarus, Barbara B., Ritter, Lisa M. and Ambrose, Susan A. *The Women's Guide to Navigating the Ph.D. in Engineering and Science.* Piscataway, NJ: IEEE Press, 2001.

Maddux, Robert B. *Effective Performance Appraisals,* 4th ed. Menlo Park, CA: Crisp Publications, 2000.

CHAPTER 6

SETTING PRIORITIES

Wish I learned this when I was a student! When I was a young professional I struggled with the differences between urgent and important, and responsibility and accountability.
            —Engineering manager at AIChE seminar

I ONCE ASKED a young civil engineer in a seminar to share with the group how he set his priorities at work. He said that he made a list of things to do every day and worked down the list, crossing off things as he accomplished them. Fine. I asked him if he ever got stuck deciding between two tasks to be done next. He replied that it happened often, so I asked him what he did then. He thought for a moment, then confessed to all (including some of his colleagues who were present) that he got up and went for a 15-minute smoking break. When the laughter died down, I persisted: what he would do when he returned from his break and the decision was still there to make? He said that he'd often just sit and stare into space, not knowing what to do next.

It *is* often very hard in our busy lives to decide what to do first. Does this sound familiar? You sit at your desk looking at piles to the left of you, piles to the right of you, phone call slips demanding return calls, your groaning paper and e-mail inboxes silently demanding your attention (my e-mail inbox usually contains around 100 messages at any given time), and your "to do" list staring at you. What's a person to do? How *do* you decide what to do first? You are not alone in this often-difficult decision we face every day in deciding on our priorities at work and in our personal lives. There's got to be a better way, and there is: learning the mundane-sounding skill of "setting priorities."

Stuff You Don't Learn in Engineering School. By Carl Selinger.
ISBN 0-471-65576-7 © 2004 the Institute of Electrical and Electronics Engineers, Inc.

> I've been so busy that I haven't really had time to sit down and think about where I'm going in my life and what my priorities are right now. I need to stop and think more about my priorities.
> —Engineering student at Tau Beta Pi seminar

But first though, just how important is "setting priorities"? Most of us rarely use this phrase in a sentence, so what's the big deal? It might, therefore, surprise you to learn that a survey of engineering managers (recounted in Appendix 1) indicated that this was an important skill for young engineers, and that they were not very good at it. Part of the reason is that engineers often don't really understand what their managers expect of them regarding setting priorities.

> I have seen "setting priorities" in practice being a problem for most people coming out of school.
> —Engineer at Rice University seminar

An illustration may help: You finish a report on Project X on a Friday afternoon and proudly bring it in to your manager, expecting thanks and praise and being told to have a great weekend! But, to your surprise, your manager says "OK, fine, but where's the update on Project X . . . that's what I really need." "Oh," you weakly reply, "I didn't know you wanted that first." In response to this, your manager may turn colors that you didn't think possible, or may just be frustrated that you didn't get clearer direction. If you only knew what the manager had wanted, you would have done that.

> I'm most concerned to ensure that my priorities are aligned with the priorities of my superiors.
> —Mechanical engineer at ASME seminar

So how do you know these things? How do you keep track of all that you have to do in the busy "real world"? And then how do you identify the most important things you have to do at any given time? Well, a first step is to regularly ask yourself those very questions.

First, how do you keep track of things? Do you make lists, like most people these days? Many people are fond of this approach, and

Figure 6.1 A list of things you need to do may work fine, as long as it doesn't take too much time to write each day.

over time adapt their lists to help them through the next hour, the day, the week, whatever. Then they try to sort the list into priorities and perhaps cross off things as they finish them. You know the drill. This may be fine, unless you spend too much time just making the list—and *you* need to define what "too much" time is—and revising it repeatedly, so you may begin to lose sight of the critical issue of actually *doing* the things. Now that you have your to-do list, how do you figure out setting priorities, that is, what is done first, second, and so forth?

ARE TASKS "IMPORTANT" OR "URGENT"?

A good way to sort your tasks is to determine if they are *"important"* and/or if they are *"urgent,"* and if they are both important *and* urgent. Things that are "important" are usually easy to identify, but they also tend to take longer to do, be more involved, and perhaps involve an unpleasant task like dealing with a problem. We thus tend to put off the "important" items with "I'll get to that later." Some examples of what usually are "important" include tackling an important work task, deciding on going to graduate school, looking for a new job, going to the doctor because you are not feeling well, getting started on a difficult/complicated task, and calling someone to deal with a serious problem. You *know* they're important, but you just have trouble getting to them.

> I need to develop my ability to prioritize. Since I like to procrastinate, everything that is "important" becomes "urgent."
> —Mechanical engineer at ASME seminar

The "urgent" items are usually the culprits that drive your priorities. You know them—the things that were "due yesterday," the "added-starters" that pop up in your inbox and have to be dealt with quickly, the report you promised you'd finish in two weeks and now "two weeks" has crept up on you, and the myriad of other items that appear on your radar screen each day. Urgent items *demand* your attention, and the important items just have to wait until when?

> **Setting Priorities**
> - What should you do first?
>   - Is it important?
>   - Is it urgent?
>   - Is it important *and* urgent?
> - Use focusing questions to help decide
>   - Which do you want done first?
>   - What do you see as the next steps?
>   - What are the critical issues?

Hopefully, it goes without saying that tasks that are both "important" and "urgent"—like my submitting the manuscript for this book to the publisher!—are your highest priorities at any given moment.

"TELESCOPING"

Returning to those who are list makers, a technique called "telescoping" (I don't know why it's called this!) may be very helpful to sort out a list of items very quickly. No, fellow engineers, I am not about to present to you an elaborate formula for weighting and ranking your priorities that some of us would find comforting!

"Telescoping" is a lot simpler and more effective. For example, take your "to-do" list of (say) 15 items and give a #1 ranking to the item that clearly is the most important or urgent. Then—and here's the kicker—*do not* try to pick #2. Instead, look for the *least* important/urgent item on the list, which is probably very easy to spot, and give that a #15. *Then go back* and look at the remaining items for the most important/urgent item, and give that a #2. Then give the remaining least important item a #14. You can sort out your "to-do" list in a few minutes by toggling back and forth like this, and give yourself enough comfort and confidence that you have set your priorities correctly.

"Telescoping" to Quickly Set Priorities
- First, pick the most important and give it a #1.
- Then pick the least important and number it as last.
- Then go back and pick the remaining most important and give it a #2.
- Then pick the remaining least important task.
- Continue toggling back and forth until you rank all tasks.

PDAs

Another new helper with our task lists are Personal Digital Assistants (PDAs, like Palm Pilots). I personally went from being a habitual list maker to a PDA user almost instantly after I found that I could categorize things to do in rough priority order; for example, assigning each task a priority number from 1 to 5. Even during the writing of this book, I've evolved the use of my PDA by adding two new categories called "This Week" and "Today," and putting task items under each category. Again, you need to find out what works for you. PDAs are certainly a most effective way today to integrate your to-do list with your calendar and your contacts, so if you have a very busy and complicated life, I urge you to see if a PDA is right for you.

It's been most helpful for me to actually make a to-do list every day. I've always had an organizer, but I never used it much. I've tried to rely on my mind more. Obviously, that wasn't working for me.
—Engineer at Tau Beta Pi seminar

Now, here are several final thoughts about the unglamorous but important task of setting priorities. First, for some of you who still try to keep all your doings in your head, I'm here to tell you that *you cannot do that anymore.* The days are over of being able to keep track of all your tasks and appointments using only the "CPU" between your ears, also known as your brain. Your life is becoming more and more complicated, and you are only setting yourself up for failure—the time when you will forget something very important and then have to deal with the consequences. You are forcing your brain to work overtime to keep track of all these things, and you may, like I remember experiencing, often wake up in the middle of the night with something on your mind—which is your brain's way of telling you not to forget that something. The minimum you need to use is a calendar of some sort, and a way to keep track of tasks with lists or a PDA.

Another thing I want to stress is that you really have to learn to actually *do* the tasks in roughly the order you have prioritized them. Duh! Otherwise, why bother taking the time to determine your priorities? But this is much easier said than done, especially if the things you rated the highest are unpleasant, take a long time, and so on, when you can dip down the list and "get rid of" the lower-rated, easier-to-do items. You'll find, like I have on many occasions, that by the time you get rid of all those easier items, either most of the day is gone or you're out of energy.

Here's what I've found helpful to focus on tackling my highest priorities. For many years on my way into work, there was a long escalator ride up from the train station into my office building on the last leg of my trip. It took about 30 seconds to go up and, in that time, I focused on the single most important thing I needed to do first when I got to the office. Just like a team huddling together before a game and shouting "Let's Go!", I psyched myself up to boot up my computer and *ATTACK that most important task* before anything else distracted me. Even despite my best intentions, I was able to do this only about two or three times out of the five-day workweek, but on this I only had to answer to myself. I have to tell you that on those days when I was able to do that important item, I felt great. In fact, I once read a management book that said if you only accomplished that single highest-priority item every day, you would be successful.

To illustrate how good you can feel, let me tell you about an e-mail I received after my seminar at Bucknell University. It seems that one of the graduate students had a part-time job, and had been stressed out about a problem with a customer, and had put off contacting him. Following my advice to act, the very next day he sucked up his courage and determination and called the customer the first thing in the morning—and the problem turned out to be no big deal and was quickly resolved. He was so relieved!

Sometimes, it is simply not clear to know what to do first. Try using what I'll call "*focusing questions*" with your manager or others to help determine priorities when it is not clear how to proceed. "*Which do you want done first?*" is a good question to ask when you have more than one task. This avoids the question to a manager: "When do you need these two tasks," and getting the answer "I need both yesterday!"—which is not helpful. Rather, asking your manager "Which do you want done first?" will, in most instances, elicit a helpful answer, and a clear direction. I have proof. One Friday afternoon, I bumped into a colleague, George, who was clearly stressed out. Apparently, he had promised Morris, one of our executives, that he would finish two separate reports by the week's end, and here we were. I replied that, actually, I was in the same situation, and had called Morris that afternoon about my dilemma, and he had given me until next week to complete one of the tasks. George was stunned. "How did you get him to agree to that?" he asked. I said I simply asked Morris which task he really needed today, and he told me, and said to get the other one to him the following week. It was as simple as that.

Another helpful focusing question is "*What do you see as the next steps?*" This question is very useful, not just in setting priorities. Try it whenever you don't know how to proceed. Perhaps you're discussing a situation or a project with your manager or a client, and there's a pause in the conversation. A way to unlock ideas is simply to say something like: "What do you see as the next steps here?" It's amazing how this can spur ideas from the other person and give you a clear window on their thinking.

A last focusing question that I would recommend you commit to memory is "*What are the critical issues?*" (I find it amusing that I learned this from a consultant whom I did not respect very much, which shows you can always learn something from anyone.) The consultant bragged to me that he met a prospective client in Europe, but had not done any "homework" on researching the company, and he just "winged it." To break the ice at the start of the meeting with the head of the company, he asked, "What critical issues does your company face?" And then he proceeded to take notes as the executive proceeded to brief the consultant on all the important problems and opportunities they faced, and he translated much of this into his work plan for a lucrative consulting assignment. But the fact remains that identifying "critical issues"—both problems and, to be sure, opportunities—is a valuable question and focusing tool for you to use in many situations.

So now I hope that "setting priorities" will become an important and urgent aspect of your skill base. You simply need to know what your priorities are at any given time, and then, of course, you need to follow through and accomplish them! Good luck!

"SETTING PRIORITIES" EXERCISE

This exercise is designed to practice setting priorities, as we covered in Chapter 6, using a hypothetical situation in which a colleague has had an injury, and the individual or a group has to handle her "to-do" list and set priorities to take action with only their understanding of each item. They need to judge each item's importance and urgency, and this should be discussed afterwards. Note that the health issue (seeing a doctor) is to be stressed as a top priority that young people often minimize the importance of.

Memo

We've just learned that one of our team members, Jane, has broken her leg skiing and will probably be out for several weeks. We need to discuss how to handle the work she is supposed to do. Fortunately, she left the following "to-do" list on her desk. We need to get back to the general manager by the end of today as to *what order of priority* we will handle these tasks, *who* will do each, and by *when.*

	Priority
• Do web research on robotics project	_____
• Order pizza for Tuesday staff meeting	_____
• Call Professor Smith about term paper	_____
• Make doctor's appt. about stomach aches	_____
• Get information about 2003 team budget	_____

- Call Joe about dinner Friday
- Finish R&D status report due Thursday
- Update resume
- Develop list of vendors for New Routes RFP
- E-mail Tech Group about Monday meeting
- Redo draft letter to Planning Board 4
- Do agenda for Tuesday project meeting
- Find out time for tonight's movie
- Research grad schools
- Call Harry about delay in estimate
- Confirm today's lunch with Ann

SUGGESTED READING

Carlson, Richard. *Don't Sweat the Small Stuff: Simple Ways to Keep the Little Things from Taking Over Your Life.* New York: Hyperion, 1997.

Mayer, Jeffrey J. *If you Haven't Got the Time to Do it Right, When Will you Find the Time to Do it Over?* New York: Simon & Schuster, 1990.

BEING EFFECTIVE AT MEETINGS

What concerns me most about the "real world' is how much time people feel they waste in meetings.
—Mechanical engineer at ASME seminar

I HAD DINNER recently with Mary, a former civil engineering student of mine from Cooper Union, and she was talking about her first full-time engineering job overseeing elevator contractors at a new office tower being constructed in Times Square in New York City. During the conversation, she told me about a difficult time that she had had at a meeting she ran that week.

It was a weekly meeting with the contractors she oversees that had, until then, been led by her manager. The previous week, Mary was told by her manager that she should now begin running the weekly meeting, and she had been very nervous. She asked her manager what she should do, and he said "Don't worry, you've been at the meetings. You don't have to do anything, they'll just start talking about things." The meeting started and no one said anything, and Mary didn't know what to say. After a long silence, she got panicky and looked over at her manager and silently mouthed "*Help!*" in front of everyone. Her manager then spoke up, said a few things, and the meeting proceeded to go well from there. Mary felt humiliated; however, she didn't say anything to anyone, and she was really afraid of running next week's meeting.

After hearing this, I asked her if she wanted me to take her through the "meetings" segment from my seminar; essentially, what you are about to read. She agreed. So, for the next 15 minutes or so, I went through the nature of meetings, how to prepare for them, who to invite, how to invite them, preparing and sending an agenda, starting on time and keeping to time, running an effective meeting, and following up to ensure action. While we chatted about her situation, I let her think about how she should apply these concepts at her next contractor's meeting, not tell her what I thought she should do.

The following week, I followed up (actually using Instant Messaging—IMing) to see how the contractors' meeting had gone, and her response was: *"Thank you! Thank you! Thank you!"* "Mary," I IMed, "you're very welcome."

The phenomenon called "meetings" is something you don't really experience in engineering school. It's not until you hit the wonderful world of work that you face what, for many, can only be called "colossal time wasters"—often sapping your energy, tyrannizing your work schedule, and frustrating you to no end. (I can almost see some of your heads nodding.) That being said, meetings can be very important and useful in accomplishing our work. I rely upon effective meetings to get people together to energize them to accomplish terrific things. So what's the contradiction here?

> Perhaps the only situation involving oral communication that people dread more than the presentation is the meeting. For many, the Dilbertian image of one person droning on endlessly while everyone else is asleep (or wishes they were) is all too real. Yet, meetings do not have to be that way.
>
> —Mechanical engineer attending ASME seminar

Before we go further, let's talk about the many different types of meetings that we attend. They come in all types and sizes, and go by names such as staff meetings, client meetings, project reviews, public hearings, and endless committee meetings. They take place in many activities, not only work but also with our clubs and other professional societies. A familiar phrase is to be "meeting-ed out," for good reason. After recently finishing my 31-year career at The Port Authority of New York and New Jersey, a friend asked me how things were going with my new business activities. I replied that, in some respects, I was involved in more things than when I was at the Port Authority. "How could that be?" he asked. I replied, "Well, I have fewer meetings that I need to go to." He nodded knowingly, with some envy.

"MEETINGS, BLOODY MEETINGS!"

Meetings become a problem—a "necessary evil"—when they are poorly conceived, badly run, and/or have no follow-up to ensure that what's been discussed or decided on actually gets done. It is almost legendary, to the point where some may be surprised to hear that the wonderful British comic actor John Cleese (of Monty Python, "A Fish Called Wanda," and Q in recent James Bond flicks) actually enjoyed making corporate training films for awhile, including his classic "*Meetings, Bloody Meetings*." (This hilarious account of Cleese playing a plant manager trying to run a meeting, and not succeeding, is probably available in libraries for borrowing and viewing.)

Meetings are commonly so awful that rarely will you attend a truly effective meeting—except for *your* meetings when you apply the guidelines set out here. Before we get to those guidelines, let me share another story that shows how seriously people view meetings.

One time during the question-and-answer period of a seminar I gave in Albany, an older engineer got up amidst about 50 people and

Figure 7.1 Will this meeting be valuable for these five engineers or a big waste of their time? Hopefully, they use certain approaches to help run an effective meeting.

angrily lit into me about my referring to meetings as "necessary evils." (I confess, I did say it.) He felt that meetings were extremely important and was outraged that I should suggest they were otherwise. (I wondered at the time what meetings *he* went to, but held my tongue!) So I replied that, OK, I agreed with him that they *can* be useful, but that they *only* become that way if you run them correctly.

So let's get at the fundamentals of an effective meeting, which are certainly easier than differential equations, but in some ways more complex too! It sorts out to things you need to do *before* the meeting, things you do *at* the meeting, and, maybe the most important, things you do *after* the meeting to make sure actions and decisions made at the meeting are actually done.

The Keys to Effective Meetings
- Meetings can be important and useful, but often are a "necessary evil."
- Don't hold a meeting unless your really need to—try calling instead.
- Meetings must have an agenda, the last item being "Review action items and next steps."
- Participate actively; if you lead the meeting, get everyone's input.
- Respect people's time: start on time, end on time.
- Send participants brief minutes/action items in 2–3 working days to avoid "meeting amnesia"!

BEFORE A MEETING

Before you do anything, you need to think hard about whether it's absolutely necessary to hold a meeting in the first place. Do you really need to invite people to come together in one place to deal with certain issues, get updates, discuss progress, and make decisions? Or can you do this by sending an e-mail to all, or making a conference call? Traditional meetings are pretty detailed undertakings, from seemingly mundane tasks like inviting people, reserving a room, arranging for support and audiovisual equipment, catering in food and, most vexing, finding common times on people's busy schedules when they can come to a meeting. All this by itself can be a nightmare for the support staff who usually sets up the meeting, with the ordeal increasing almost exponentially as more people need to be invited. So, only hold a meeting if there is no other way.

MEETINGS MUST HAVE AN AGENDA

The first requirement for an effective meeting is that there *must* be an agenda. How important do I think this is? I recently interviewed finalists for a position and, when asked how they would deal with the difficult people and the meetings that would be involved, the person who got the job was the only one to say immediately, "First, you need an agenda."

The agenda is, quite simply, a road map for the meeting. If you don't have one, you risk going all over the place. People like to talk, in case you haven't noticed by now, and you won't get the things discussed that are the purpose of the meeting.

> You must use an agenda during meetings. Most meetings I attend are for topics with no agendas.
> —Mechanical engineer at ASME seminar

So you should expect that any meeting that you go to or, of course, that you run, will have an agenda. If you've been invited to a meeting, ask to be sent one, or say you won't go! Why would you think of doing this? Because you're a busy person, and you're not sure if you will be needed there until you see what items are to be discussed and what information you will be asked to cover. Please don't make the support person who's only trying to arrange the meeting uncomfortable, but ask that the person whose meeting it is to get back to you.

> [T]hese are practices for good meetings, especially asking why should I go and what do I need to contribute.
> —Engineer at major Midwestern engineering company

To be sure, younger engineers often don't have much say as to whether or not to attend a meeting. Many is the time I've heard attendees say that they don't know why they're there—"I was told to come." And they also don't often get a chance to arrange or run meetings, but they *can* participate effectively by raising points during the conversation and asking questions, and also by observing how well or poorly the meeting is run for future reference.

DEVELOPING AN AGENDA

Identify the topics that need to be covered at the meeting. Think of what needs to be decided on, what is to be presented for information, and what items need to be reviewed and discussed. Each agenda item should have a person named to lead the discussion, and the approximate time to be spent on that item should also be indicated, given the total time allotted for the meeting. The first agenda item should always be to take a moment to allow everyone to introduce themselves, since it's rare that everyone knows everyone else, and to review the agenda. Don't hesitate to reorder items if necessary, adjust times, and so forth.

Think about the order of the agenda items, and try to deal with the most important items, say, decisions, early in the meeting while everyone is fresh and to ensure that you have enough time to make the decision. Or, conversely, it may be better to start off with items that will go more quickly in order to make sure that all will be present for the discussion on the important items.

The last agenda item should always be "Review decisions made and action items." This will serve to review what important items were resolved at the meeting, and allow you to discuss them further if there was some confusion or disagreement. Don't wait to find out that there's a problem or misunderstanding until after everyone leaves. Send out the agenda well before the meeting, and certainly no later than 3–4 working days beforehand, to allow people to prepare for the issues to be discussed.

PRIOR TO THE MEETING

If you are running the meeting, it is good to send out short meeting reminders a day before or a few hours before, or call around to attendees, telling them that you plan to start on time. If you are just attending, it is a good habit to get in touch the day before the meeting to confirm that it's still on; you may not have heard of any postponement or cancellation.

Check the meeting room if you're not familiar with the layout, to make sure of seating arrangements, audio-visual equipment, and other things needed. If you are traveling to the meeting, it is good practice to get there at least ten minutes beforehand, get a seat, and put your things at your place.

AT THE MEETING

It is very important that you respect people's time, so do your best to *start on time and end on time.* It is almost legendary that meetings rarely start on time, so get the reputation that the meetings that you run start on time by doing it. Then, people will know the next time that if they walk in late, they'll have to endure the peer pressure directed at latecomers.

I once ran a meeting in the Port Authority's World Trade Center headquarters to which airline representatives were invited from the three area airports, but were reluctant to come into the city and "kill the whole day." I assured them that the meeting would start at 10 AM sharp, and be over by noon, after which they would be free to stay for lunch or go back to their airport. At 10 AM, everyone was there and the meeting started exactly on time. Do I need to say that the meeting ended at noon? One of the airline managers came up to me at the end of the meeting and said that that was the first time he was ever at a Port Authority meeting that started on time.

But, of course, the purpose of the meeting is not just to be there, but to share information, discuss issues, make decisions, and pursue actions. That's why it is so crucial to have and actually stick to the agenda. Further, each attendee should try to participate actively; otherwise, why are they there? And the chair of the meeting should ensure that everyone gives input. If you lead the meeting,

Figure 7.2   Each person should try to participate actively at a meeting; otherwise, why are they there?

you should control and guide the discussion. Don't be shy about keeping on track and not letting the conversation wander. People are depending upon you to run an effective meeting.

I once chaired a committee trying to select the best proposals from two dozen submitted in a Request for Proposals for Internet kiosks at airports. We had gotten many more responses than we had expected, and our committee was a bit overwhelmed in trying to reduce the number of responses to a manageable number of about 4–6 vendors to invite in as finalists for interviews. We were on a short time frame and had already set the dates for the interviews, which required travel for most, so it was imperative that we decide on the finalists at that meeting. But we were stalled after an hour's discussion, and didn't want to extend the meeting if at all possible, since people had other commitments.

I was chairing this meeting, and I didn't know what to do next. Everyone was quiet, thinking. Then I noticed that one of our committee members, Chris, hadn't said anything up to that point, and so I turned to her and said: "Chris, You've been pretty quiet. What are you thinking?" Well, Chris *had* been thinking, and had been a bit timid in speaking up, and she said "What if we. . . ." Her approach was the breakthrough to resolving the issue, but she wouldn't have said anything had not the chair gotten her to share her thoughts. We then finished up and ended right on time. Which brings up a second point. Another person at that meeting, Paul, was so enthused about how great the meeting went, that he blurted out to me at the end: "Carl, you must come to my department and tell us how to run a meeting. We have this big sign on the wall of our conference room about how to run meetings, but we never follow it. Would you come and talk to us?" Although it might have been easy to just say "yes," my answer to Paul was "no"; I only used the simple guidelines that most people know in running an effective meeting, which Paul himself already knew (and which are presented here). So just do it! Paul nodded knowingly and agreed.

But we're not done yet, since perhaps the most important things are done *after* the meeting.

AFTER THE MEETING

You need to *prepare and send meeting minutes quickly.* Make the minutes brief, or just limit them to decisions made and action items;

for example, who agreed to do what and by when. Don't try to write literal accounts of everything that's said; only capture key points and any decisions and actions. Record the meeting on a tape recorder if a meeting record is really needed, and end the minutes/action items with the phrase, "If anyone's understanding of the above items differs, please contact me as soon as possible, and by no later than 5 days after the meeting."

It is very important to send out the minutes/action items in 2–3 working days, no later. Why? There are several reasons. First, it informs people who couldn't attend about what took place. Second, it avoids the "illness" called "meeting amnesia," a condition that curiously affects certain people who conveniently forget that they agreed to do something at the meeting, or claim to be unclear about what they're supposed to do.

> If everyone followed this advice, meetings would be much more useful and less painful.
> —Engineer at major Midwestern engineering company

Let's end with another story of a meeting that stressed a young engineer to the bone. At seminars, I ask engineers to talk about the "worst" meeting they had gone to. One time, an engineer I'll call John started shaking his head. I immediately called upon him and he shared a terrible story. He was a junior engineer on a project in a consulting engineering firm, and a major client was coming in for a Friday meeting to review the status of a project, which was having big problems. John's manager asked him to take notes during the meeting, which he characterized as an all-morning "yelling match" that tried to resolve some critical problems. Finally, they did begin to settle their differences, and all went out to lunch afterward to celebrate. After lunch, John's manager asked him to "write up" the minutes of the meeting. John didn't tell him that his notes were a jumble of total confusion. Instead, he tried to start piecing together key points, but then gave up and decided to do it Monday when he'd feel fresher. But all weekend he was totally stressed out, only thinking about what a mess the meeting was and how was he going to unravel all the key points and decisions. When he came to work Monday, he was a total blank. Fortunately, he had a sympathetic manager who understood the situation and helped John. John should have been keeping track of key points and decisions at the meeting, not attempting to record everything, and then these key points should have been reviewed at the end of the meeting. John could then write out a list of these items.

So now you know how to run an effective meeting and participate actively in meetings you're invited to. You cannot underestimate the importance of this skill to being more effective in your career. Try these approaches at your upcoming meetings and see the difference!

SAMPLE MEETING AGENDA

As an illustration, here is an actual agenda for a two-hour meeting. It was sent to all invitees about a week before the meeting to alert them to subjects to be discussed. Note that the person who will lead the discussion on each item is indicated as well as a rough time allotment, and that the last item is "Review action items and next steps." Even though this was a very ambitious agenda, the meeting adjourned at 8 pm.

Cooper Union Alumni Association
Executive Committee Meeting Agenda
Thursday, 9/23/03, 6-8 pm, Alumnispace

Tentative Agenda (Person responsible and approximate time allotted)

1. *Call to order*—promptly at 6 pm.
2. *President's Report* (Carl) (20 minutes)
   a. Goals: Strengthen CUAA; engage all alumni; achieve Annual Fund goal
   b. Review CUAA action items (see attachment)
   c. Selection process for new Associate Director of Alumni Relations
   d. Website renewal put on month-to-month (Yash)
   e. Wine labeling concept (David)
   f. Update on Alumni Directory Project (David)
3. *Annual Fund report* (Stacy) (10 minutes) (see attachment)
   a. Status
   b. Strategies to achieve higher giving

4. *Trustees Report* (Carl/Carmi/Sue) (20 minutes)
   a. Review issues at Trustees meeting
      i. General tenor of Board, financials, programs, building, capital campaign, engineering curriculum
   b. Importance of alumni trustees attending Executive Committee meetings
5. *Secretary-Treasurer report* (Yash) (10 minutes)
   a. Alumni presentation to freshman orientation
   b. All-student cruise attracted 400+ students!
6. *VP–Faculty Liaison report* (Matthew) (10 minutes)
7. *Committee updates* (20 minutes)
   a. Events Committee consolidated (Yash/Miriam/Rebecca/Margaret)
      i. Art Auction/Casino Night (10/4)
      ii. On the Rooftop event for young alumni (10/4)
      iii. Added starter: Fall event at DIA Museum Beacon (10/19)
      iv. Merged committee with Student Events and Young Alumni
      v. Prepared unified event calendar
   b. Constitution Committee status and process (Kim)
8. *VP–Membership Report* (Kim) (15 minutes)
   a. Update on Student Ambassadors
   b. Review Class Rep action strategy
9. *Decide agenda for 10/14 Council meeting* (Carl/David) (10 minutes)
10. *Review action items and next steps* (Carl) (5 minutes)
11. *Adjourn* (by 8 pm)

SUGGESTED READING

Doyle, Michael, and Straus, David. *How To Make Meetings Work: The New Interaction Method.* New York: Jove Books, 1982.

CHAPTER 8

UNDERSTANDING YOURSELF AND OTHERS

I have an inability to get started and/or complete projects when I don't see any relevance for them. I also have difficulty starting new things.

—Engineering graduate student at MIT

IN THE EARLY years of my career, I never understood when people would say that they were trying to "find themselves." It took me well over 20 years to finally appreciate the importance of understanding your own self—your individual personality, and what things you can do to develop your personal skills and abilities to synchronize with your very personal needs and wants. *The better you understand yourself, the better you will then be able to understand and appreciate other people with whom you will work and live during your life.* Some areas where this will be of particular value to you are in asserting your ideas, improving your thinking process, communicating well with your manager, participating more effectively in meetings, and dealing better with people and avoiding conflict.

I believe there are several reasons why engineers don't develop good "people skills," even when it comes to understanding their own selves. These reasons include a tendency to become an engineer in the first place because we are more comfortable working with numbers and scientific principles, and don't have to deal as much with other people. This may be an overstatement or too broad a generalization, but it seems to fit. Engineering school only reinforces this through the technical curriculum that crowds out all but the most minimal humanities coursework, and minimizes opportunities to work together, except on technical project/design teams.

Stuff You Don't Learn in Engineering School. By Carl Selinger.
ISBN 0-471-65576-7 © 2004 the Institute of Electrical and Electronics Engineers, Inc.

When you emerge from engineering school and start your career, there continue to be relatively few opportunities or even the desire to work with others—certainly compared with other professions—since your technical abilities can be applied by working by yourself in many instances. Yes, I know there are plenty of opportunities to work with others in your engineering careers, but I believe that many/most engineers are predisposed to working alone. There is even an unflattering term for it, "a skunk works," the somewhat derogatory but complimentary reference to the brilliant techies in the back room, but don't let anyone see them, certainly not our clients/customers!

Another reason that I feel that many young engineers don't focus on their personal development in the early years of their careers is that they have been on a comfortable track to be an "engineer" for many years. We were good in math in high school, got into and through engineering school, and got a good entry-level position employing the latest technology skills, all the time perhaps not questioning nor aware of larger personal concerns about how well suited we are to this career and some things that may be bothering us.

> What concerns me the most is how to deal with professional people and have to prove to myself that I'm really on the right path doing the right thing in the right place.
> —Young engineer at AIChE seminar

Let's not go any farther here, even though I've raised a number of provocative and arguable thoughts, and hopefully have been honest but not too negative, since there are so many personal variables with each of us and dynamic changes in the engineering field. Rather, this chapter will share with you some things I've learned that will help you now to feel better about yourself and enable you to work more effectively with others.

BE ASSERTIVE—NOT PASSIVE OR AGGRESSIVE

As I've said elsewhere in this book, *you* have the responsibility to control your behavior and how you act in the real world. Perhaps nowhere is this as important as how you deal with others, and being "assertive" is a crucially important skill. But this is actually very hard for many people, not just engineers. So what are we talking about here? First, it is very hard for a young person (or anyone, for that matter) to "speak their mind" in the workplace. Despite all your learning and skills, you have entered a work world where you are a "rookie"—an inexperienced person who may not often be called upon to give advice or tell people what you think about things. This is because you haven't yet developed a "track record" of specific accomplishment, and those who are older and more experienced—even your manager—can't or won't respect your abilities, or are reflecting a natural "generation gap," however unfair that may be. Coupled with this is the fact that many engineering students have been loath to speak up in school. Very few courses or professors will try to draw out the quiet students, who feel they can "get by" and never open their mouths, as long as they get the work done and pass the course.

> I'm concerned that I am too self-involved in my work—take things too personally. Also, it's very difficult to deal with management that doesn't seem to listen.
> —Mechanical engineer at ASME seminar

So here you are at work and perhaps frustrated at not being able to get your ideas across, or just uncomfortable that no one is asking what you "think" about something. This is not good. You need to be able to feel comfortable asserting yourself, and by this I do not mean getting people angry with you for speaking out of turn.

If you think you are the only one who needs to be more assertive, you are wrong. There are many titles at your local bookstore on "assertiveness," and it is a typical course in many "continuing education" programs. In essence, to be assertive means to say something in a direct, positive, and nonthreatening way. Being assertive is somewhere between being "passive," the unconscious mode of choice for many young engineers, and being "aggressive," which is an unattractive and usually ineffective way of forcing your thoughts on others, for example, "getting in their face."

Let's illustrate with a mundane but common situation—deciding where to go for lunch with someone. The person you're going to have lunch with asks you, "Where do you want to go for lunch?" If you're passive, you reply with something like: "Oh, anywhere you want to go is fine with me." Or, "I don't know, you pick someplace."

If you're aggressive, you might reply: "I only want Chinese food. If you don't want that, I'm not going." Or "I'm picking a restaurant this time. The last time, the place you took me to was terrible."

If you're an assertive person, you might respond in these ways: "Since the last time we went for Mexican food, how about we try that new Chinese place over on Main?" Or, "Well, I'm not sure, but I really feel like Chinese food today, so pick a place." Or, "Let's go someplace close by since I only have an hour for lunch today."

I hope this illustrates the differences between passiveness, assertiveness and aggressiveness, so that we don't have to go to a dictionary for further help. Now let's talk about some work situations and how you can be more assertive in your behavior. Such situations occur all the time—in conversations with your manager about your work, in dealings with clients or customers, and in participating at meetings. The point is that for you to be more effective in these situations. You need to learn to be assertive, so force yourself to get more involved if you tend to be passive, and tone down your approach if you are on the aggressive side. Given that recommended approach, here's how to be more assertive in certain situations.

MEETING SOMEONE

An important time to be assertive is when we meet someone and, as common in Western culture, shake hands. Whether this is between old friends, meeting someone new for the first time, or saying hello at a job interview, the handshake is a way of asserting yourself positively with the other person. Done well, it tells the other person that you're glad to meet them, and look forward to talking with them; done poorly, it may signal that you are a weak person or, the other extreme, too aggressive for their taste. How can all these feelings be transmitted in a simple handshake? Well, first let's describe the proper way to shake hands. When you approach someone and extend your hand, it is important to make eye contact with the other person. Looking away, especially looking down, could signal that you're afraid of them or their authority. And don't forget to smile, to tell them that you're happy to see them (even if sometimes you're not).

Another aspect of the handshake is how firmly you should grasp their hand. A rule of thumb (pun intended!) is to shake some-

Figure 8.1   Make eye contact when shaking the other person's hand. It also helps to smile! Shake hands as firmly as the other person grasps yours, and hold the handshake for as long as the other person does.

one's hand as firmly as they shake yours. Important: Don't just hold your hand out limply for the other person to shake (a sign of passiveness), and, on the other extreme, don't grab their hand and squeeze so hard that you hurt them while also perhaps pumping their hand up and down (an aggressive approach). How long you should shake hands? Probably as long as the other person wants to, and then quickly release your grasp. If you are concerned that you're not understanding all of these tips, here's an easy suggestion: practice with someone! Get off to the right start with someone with this important ritual in interpersonal relations.

RESPONDING TO "WHAT DO YOU THINK ABOUT THIS?"

Perhaps no question unnerved me more as a young engineer than being asked what I thought. I didn't know what to think! In engineering school, no one ever asked me what I thought, certainly not most professors. I was in the business of trying to learn complex technologies, getting the right answers from formulas, and, mainly, passing courses. There didn't seem to be much to "think" about.

Cut to our real world, and real projects on your plate. You now are *expected* to give your thoughts on many things. And it is not acceptable to simply shrug your shoulders or give a blank look and say "I don't know."

Important digression: "I don't know" is a perfectly acceptable response that you will use many times in your career. Say it if you really don't know something, and, if appropriate, add that you'll find out and get back to that person. Don't think that you have to know everything; no one knows everything, or anywhere close to it, in our complex world. In fact, there are only two categories of things that you should know: (1) What you know in your head, and (2) Where to find out the rest.

Back to "What do you think?" Here's a way to help your thinking process. *Before anyone asks you,* get in the habit of mentally answering the question of what you think so that you have a ready, thought-out answer. For example, what's your answer to the following question from your manager: "What do you think we ought to do next on your project?" Think about this a moment, and say aloud your response. (OK, you may say this silently if you're among people while reading this!) Get in the practice of thinking about things in a structured way. By the way, here's an improper but tempting answer to your manager's question about what you think: "Gee, I don't know what to think. You're the manager . . . you're paid lots more than I am . . . so you tell me!"

Listen up: if you don't assert yourself and, for this example, don't say what you think the next steps might be, then you're as much telling your manager that you are clueless and unable to contribute your ideas to helping your employer.

Some Keys to Understanding Yourself and Others
- Be assertive, not passive or aggressive.
- Understand yourself better and how to relate to others . . .
  - o Take the Myers–Briggs Type Indicator (MBTI)
  - o What do you *like* to do?
  - o What *don't* you like to do?
  - o What about *others*? (Ask them!)
- Updated Golden Rule: "Do unto others as *they* would have done unto them."
- Determine your mix among career objectives: money, job security, challenging job, recognition.

How else can you practice being more assertive? Here's another easy way: read editorials and "Op-Ed" essays in newspapers and magazines. Editorials are the opinion of the publication on all types of subjects. Many newspapers have short "Op-Ed" articles, which stand for "opposite the editorials." Perhaps you all know what these are, but I'll bet that many of you have never read them or don't read them regularly.

How can reading editorials and Op-Eds help you be more assertive? Because they are written on subjects on which the writers assert their opinions on specific issues, provide their explanation of the issues and options to deal with it, and offer their recommended approaches or positions, that is, "what they think." You will see how the writer is assertive, in addition to being more informed about the issue itself, and this will help you see how to present information and thoughts in ways that convince others of their merit.

To help engineering students understand assertive writing, I give my Cooper Union students a homework assignment to read a *New York Times* editorial that covers "truck sizes and weights." They are asked to: (1) Summarize the editorial in *one* sentence; (2) Say *whether they agree or disagree* with the editorial's position and why—"Yes" or "No" are acceptable; "Maybe" or "It depends" or "I don't know" are not allowed; (3) Identify the strongest and the weakest points supporting the position; and (4) State how they would change the editorial to make it more influential.

SPEAKING UP AT A MEETING

We talked a lot about meetings in Chapter 7, but let us further discuss here about how to be more assertive at a meeting. First, we need to understand that young engineers are not likely to feel comfortable saying *anything* at a meeting, nor are they often called upon to participate actively. But you are there and, other than listening and absorbing information, you *should* feel comfortable contributing where you can. In fact, I would recommend that you open your mouth at least once during the meeting, besides introducing yourself! How do you do this?

- Ask a question. ("Joe, Where do you think that issue will affect my project?")
- Ask someone to clarify a point you didn't understand. ("Sue, would you please go over that point again, I didn't quite get it.")
- Bring up a point that you think will be helpful. ("Frank, perhaps you already know this, but they're doing a project like this in England, and I have some information on it.")
- And, when you really feel comfortable, offer your thoughts or suggestions! ("That's an interesting point. I think we should consider doing thus-and-so.")

No one is going to bite your head off if you participate at a meeting. On the contrary, others may be pleasantly surprised that you spoke up, and people may come over to you after the meeting wanting to know more about what you said. For example, I was startled one time to get a handwritten "Thank You" note from my Department Director after a briefing for my agency's Executive Director on an important project; he congratulated me on sharing my thoughts, even though I just had blurted out one fact spontaneously during the meeting. So assert yourself; don't just sit there!

> I have to learn to speak up and ask questions when I need help—I can't worry about what someone else will think.
> —Engineer at Tau Beta Pi seminar

DISAGREEING WITH SOMEONE AND AVOIDING A CONFLICT

Being assertive is a crucial skill to employ when you have a difference of opinion with someone, which can range from polite disagreement to open conflict. One assertive response to a statement you disagree with can be "Let me see if I understand what you've just said. You said that blah-blah-blah." Repeating or rephrasing what the person said might actually clarify the point so that you now agree with it, or find out that you misunderstood what they were saying. Or you can clarify that you still disagree.

The next level of assertiveness might then be, "*What are you trying to accomplish by doing that?*" (By the way, this is a very useful question in many contexts, e.g., when you are unsure how to respond to something.) Again, perhaps this question will help clarify the goals of the other person, which you may agree with. Further, after hearing the response, you can share this with any others: "Do others agree with this?" or "What do others think about this?" I once used this response in a meeting with a local community planning board in New York City when one of the attendees suggested a revised subway routing that seemed outlandish to me, but workable to him. Stumped for a positive response, I stumbled on asking others what they thought of the idea, and (to my pleasant surprise) there was unanimous outcry against it from the others.

You can't always avoid disagreements in our complex world. People have different opinions and agendas, so sometimes you have to be assertive in seeking closure or finding someone to resolve a dispute. An assertive way to respond when you reach an impasse with someone may be to say that "we agree to disagree." If the situation needs to be resolved, then you have to find an appropriate person or entity that can decide, usually a person at a next-higher level in your organization.

Here's another way to assertively solve a conflict. One time I received a phone call from a colleague who was very concerned about something I had done. It's so long ago now that I have no recollection of what the subject was, except that I had certainly made a big mistake. However, I clearly recall asking him during the phone call: "Dick, are you angry with me?" I think Dick was a bit startled with such an open question, and, after a pause, said, "Well, yes, I am angry at you!" I then offered to go out to his office that same day to discuss the situation, and he agreed; we resolved the issue and became closer friends afterward. So don't be afraid to assert yourself, it will make you more effective in many situations.

SMILE MORE!

I wasn't sure where to position this important subject, but this is a good place to do it. Part of dealing effectively with other people is relating to them as human beings. And when human beings are happy, the universal expression is the smile. So just because we're engineers doesn't mean we have to be serious all the time, with a stone face that hides whatever thoughts and feelings we may be harboring. Take a look at Figure 8.2 and see how much more appealing you can be when you smile. Which person would you more like to deal with? So smile when you first meet someone, during your conversation from time to time, and when you're speaking to an audience. More people may like you just for that.

UNDERSTANDING YOURSELF AND OTHERS

What makes us tick? We are complex animals and, yes, we *are* animals, so there are many levels to begin understanding ourselves.

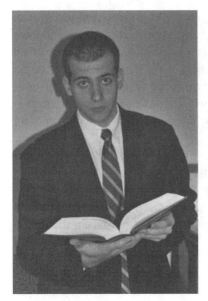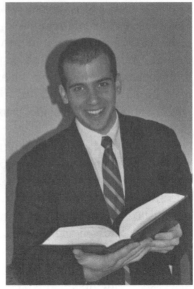

Figure 8.2   Who says you have to look serious all the time? (And what are you thinking?) See how a smile makes this person more appealing and likeable?

What drives us, what stresses us, why do we behave as we do? Let's again try to sort this global and cosmic area into some specific ways to understand ourselves and then others, and how to relate better to other people.

It is so important to first understand yourself as a baseline in order to relate better to others—coworkers, managers, clients, friends, and family. Too bad we can't just plug ourselves into a diagnostic machine like we now use to troubleshoot the mechanical status of our car. (Maybe we will be able to do this in the future!) Each of us cannot tell another person how we *really* feel; we can only use words to express happiness, sadness, anger, stress, disappointment, and so on. We can only show how we feel in our behavior and in our words. So where should we begin to better understand ourselves?

A good way to start understanding your own personality is to read the book, *Please Understand Me II,* by David Keirsey (see the suggested readings at the end of the chapter), which is an application of the Myers–Briggs Type Indicator (MBTI). In fact, I will say that I have taken no more effective training then this, in terms of its usefulness in assessing my own self, and in improving my relationships with others, especially with people who I had great difficulty working with.

Without attempting to thoroughly explain this (for that, please read the book), the psychologists Myers and Briggs developed a structure of 16 different types of personality that everyone fits into, based on principles of psychology. At the risk of oversimplification, they hypothesized that people have four parameters of personality preferences, ranging from being Extroverted (E) to being Introverted (I); preferring to be Intuitive (N) or leaning more toward Sensing (S); being more Thoughtful (T) rather than being more Feeling (F); and being more Perceptive (P) or tending to be Judgmental (J). Using a self-administered questionnaire contained in the book, you can determine your "type" in less than an hour (mine is ENTP), and then the fun and insight begin. Each of the sixteen types (permutations of the four parameters) has an associated profile one can read—very much akin to a "scouting report" on someone, which you might read to know what opposing players are like, their strengths and weaknesses. At first, it is startling to most people taking MBTI to see many of their inner feelings revealed in print! "How could they know?" "Gee, I didn't even know that myself, but now I see it." These were actually my reactions right after taking the MBTI course.

> How can I not be a perfectionist?!
> —Engineering student at Tau Beta Pi seminar

Before you dismiss this as "mumbo-jumbo," let me give you an example of how I *applied* this knowledge soon after receiving it. I was to brief our deputy director about the status of a new airport access service that we were heavily subsidizing until it became profitable. My manager, Sue, and I were to brief Morris, who was very concerned with the viability of the service. Now, a bit of background: Morris and I had a very uncomfortable relationship. He was like a "father figure" for me and very paternalistic toward me, quick to frown at me when I said things he didn't like, though he seemed to like me personally. So it was a bit of a mystery. In preparing for the briefing, it occurred to me to see if MBTI could help. I was able to figure out Morris's personality "type" in my head since I knew him very well, and then I read his personality profile. It was a revelation to me that people like Morris wanted the "facts" (he was an "S" person), whereas I was very intuitive ("N") and would invariably start with "Morris, things are going well . . . ," which would only make an "S" person like Morris quickly assume that things were *not* going well. So, after reading "the book" on Morris, I wrote down every fact or number that I could find on the status of the airport shuttle service; for example, how many people rode it last week, how many runs were made, revenues last period versus previous period, amount of subsidy and trend, and so on. Then Sue and I went to the meeting.

On the way to the meeting, Sue—who I had not told that I had prepared differently—told me to be brief, since my "rep" (reputation) was that I talked a lot. We got to Morris's office, and his secretary said we should go right in. Morris greeted us and said "What have you got for me?" Sue said I would give him a status report and be brief. I immediately started reciting the facts I had assembled, and Morris listened. I went through about a half dozen facts, and he asked a question and I answered. After another minute or so, I was finished with my facts and stopped. Morris then asked another question and we discussed that issue. Then Morris said "Thanks!" Only about 10 minutes had elapsed. Sue broke out of a reverie (she had been thinking about something else) and said "What? Morris, Are you OK on this?" Morris said "Yes, thanks." Sue said: "You mean we're done?" Morris again said yes and thanks. Sue grabbed my arm and said "OK, Carl, let's go!" We walked out of Morris's office and his secretary, unaccustomed to such short meetings, said "Oh, do we have to re-set the meeting?" Sue replied,: "No, we had the meeting." The secretary had an amazed look as we walked away.

After getting away from Morris's office, Sue stopped me and said: "Carl, what happened in there?" I told her how I had prepared, using the Myers–Briggs technique to get an insight into Morris's personality type, and then had briefed him in that different way. Sue suddenly beamed and said: "Wow. That stuff really works!"

I hope that anecdote shows the power of understanding how to deal with another person. This does really work, because it helps us to understand what goes on in our heads—our heads and other people's heads—and allows us to modify our behavior to be more effective in getting our ideas across, or at least reduce the obstacles. "Please Understand Me" is a very appropriate title. I have regularly used MBTI techniques to guide interactions with people over the years. Needless to say, my business relationship with Morris significantly improved after that meeting, and he may never figure out why unless he reads this book!

> No matter where you work or wherever you are, people are the same and we deal with similar situations. It is a matter of how we deal with them and what type of attitude we start with.
> —Mechanical engineer at ASME seminar

WHAT DO YOU LIKE TO DO? WHAT DON'T YOU LIKE TO DO?

Next, let's turn to what sounds like two easy questions—What do you like to do? What don't you like to do? But have you ever really stopped to think about the answers? Doing things that you like to do is an important part of your life, whether in your work or in your personal life. At the same time, knowing this does not mean that you'll only get to do the things that interest you. The real world will require that you do many things that are necessary to get a job done. After all, it is called "work" for a good reason. Yet each of us has specific preferences and dislikes that we use to consciously guide us as we choose jobs or tasks, to the extent we can.

Frankly, I had not thought much about this throughout most of my career. It wasn't until I began to feel a bit disillusioned and unhappy with some things I was asked to do at work. It dawned on me that I was unhappy in some situations that I probably had some control over, so I began to think about my personal likes and dislikes. I'll share with you a few things I came up with, which are not meant to guide you in thinking about yourself, since we are all infinitely different in our personalities and situations. Anyhow, I determined (after much thought) that what I liked to do best was work with new ideas, collaborate and interact with interesting and optimistic high-energy people, and travel more. I isolated the things I disliked: doing repetitive tasks, working alone, and repeating the same project after having done it once. These likes/dislikes actively guide my decisions on my general career objectives as well as specific tasks I pursue, and are always subject to modification and updating upon further reflection. The point is: if I am doing what I like to do, and minimizing what I dislike, I will be more motivated to pursue and be effective in my activities.

HOW TO MOTIVATE OTHERS

Now that we've talked about understanding ourselves better, let's turn to motivating others. First, *do not assume that the people you work with and otherwise interact with during your work and personal life think or feel the same way that you do.* Even if you share some common situations like working in the same organization, or are in the same family, and so forth, all of us have a different set of likes and dislikes. This can be crucially important where there are differences in attitudes, and this may not be apparent until there is some conflict.

> The fact that most engineers experience the same problems—no matter the company and that the way business is run today—is the way it is.
> —Mechanical engineer at ASME seminar

For example, you will have a great deal of interaction with your manager. Usually, this person will be older, more experienced, and have many differences from you in personal background and characteristics. Before going further, I happen to think that these differences in our increasingly diverse society can be very positive, and serve to enrich and strengthen our activities and accomplishments. However, there is also increasing room for misunderstanding one another. So where is this leading? Simply that if you don't appear to understand how the other person is relating to you, ask them about the situation. Don't be shy—remember you've just learned how to be assertive! So ask your manager to explain things if you are not clear about what is expected from you, or how you can do better. In fact, that's the way to couch it: "How am I doing? Where can I improve?" These questions allow a positive and constructive response.

UPDATED GOLDEN RULE

I think that the venerable "Golden Rule"—"Do unto others as you would have others do unto you"—is not as helpful as it once was, although it still seems perfectly acceptable that your behavior toward others might be guided by what *you* would prefer. But in our diverse global society, where we have to work with people from all over the world, others may have very different ideas about how they want to be treated, in both big and little ways. So an updated Golden Rule seems more appropriate to guide us these days: "Do unto others as *they* would have done unto *them*."

Many times, though, we don't know *how* others would like to be treated. So what should we do about this? Actually, this is pretty easy: *ask them*! You can ask them directly in conversation. You can observe people at meetings or in social situations and mimic their habits and behavior ("When in Rome, do as the Romans do"). You can read about the habits and customs of different types of people, and understand how their cultural habits differs from your own paradigm. A single example that illustrates this is that the exchange of business cards when meeting with Japanese persons is an important ritual for them. While Americans are casual about exchanging cards, the process with the Japanese is to formally hold your business card with *two* hands, *bow slightly* as you offer your card to the other, then immediately *read* the other person's card after receiving it, noting the title of the person. Understanding others is an important area for you to continuously learn over your career in order to develop the robust skills needed to relate effectively with other people.

> There is no sure-fire method for successful communication. Each of us has to discover what works for us. The only way to discover what works is to interact with people.
> —Mechanical engineer at ASME Hartford seminar

YOUR CAREER OBJECTIVES

Let's close this chapter on understanding yourself with a somewhat unusual ending—how this all fits with your career. I recall receiving a "career survey" early in my career from a professional society that was collecting information on various items—salary levels, responsibilities of engineers in certain position titles, policies of the employer on benefits, and so on. But one question really stumped me: "What's your mix among career objectives: Compensation, Job security, Challenging job, and Recognition?" The survey asked members to allocate 100 points among the four objectives, meaning that if getting the most salary was your only objective, you would give it all 100 points. If all four were equally important, you would give each 25 points.

What's your mix among career objectives? Allocate 100 points among the following objectives:

Compensation		_____
Job security		_____
Challenging job		_____
Recognition		_____
Total points	=	100

I remember staring at the survey for a long time and trying, for the first time, to wrestle with this important question. I found that the question went right to the heart of what things are important to us in our careers, and relate to our feelings and attitudes and self-esteem. So it was not simply thinking about the trade-offs, say between higher pay or increased job security. As we close this chapter, I'll ask you to think about how you'd allocate the 100 points to those four objectives. Your answer may help guide your career and personal life in a positive direction.

SUGGESTED READING

Davidson, Jeff. *The Complete Idiot's Guide to Assertiveness.* Indianapolis, Indiana: Alpha Books, 1997.

Keirsey, David, and Bates, Marilyn. *Please Understand Me: Character & Temperament Types,* 4th ed. Del Mar, California: Prometheus Nemesis, 1984.

Keirsey, David, and Bates, Marilyn. *Please Understand Me II: Temperament, Character, Intelligence.* Del Mar, California: Prometheus Nemesis, 1998.

Morris, Desmond. *The Naked Ape: A Zoologist's Study of the Human Animal.* New York: McGraw-Hill, 1967.

CHAPTER 9

WORKING IN TEAMS

I need to be able to explain to other people about the basic principles of teamwork. Some people around here do not know how to involve you in the big picture.
—Engineer at a major Midwestern engineering company

I ONCE WENT to a luncheon of a prestigious business group that attracts senior executives, and sat at a round table where everyone was a stranger. The normal "drill" in this situation is to say hello to the person on one side of you, perhaps exchange business cards and have a quick chat to see what that person is doing, and then, after a time, turn to face the person on your other side and do the same. Well, I had finished chatting with the first person, and turned to talk to the other person. He introduced himself as an executive of a major company and, when I asked what he did there, he replied sharply that he was just made "a team leader and he had no %#&$ idea what that meant." No, that wasn't a misspelling that my spell-checker missed; he actually used an adjective that I won't repeat here, but you may guess. I was startled at how he was venting his anger over this situation to someone he didn't know, and showed me how difficult it can be to change some of the rules in the workplace to effectively establish and work in teams.

Working with other people in teams, whether as a team leader or a member of the team, can be a very rewarding and exciting experience, or it can be a difficult and frustrating ordeal. The difference lies in understanding some important aspects of teams that are obvious and some that are subtle.

So why bother at all with team efforts? After all, isn't an organization by its very name—organization—designed to accomplish the work at hand with a sufficient amount of resources? You are

Stuff You Don't Learn in Engineering School. By Carl Selinger.
ISBN 0-471-65576-7 © 2004 the Institute of Electrical and Electronics Engineers. Inc.

Figure 9.1 Teams can tackle complex projects and have fun doing it, but they must follow some important guidelines to be effective.

probably a member of a unit, section, or division of an organization, and that is your normal mode or framework of your work tasks. So what are teams and why use them? Quite often the nature of special projects or efforts lend themselves to establishing a team effort to accomplish the work. This could be because it is a short-term project that needs special attention from various disciplines or skills, or it could be a longer-term project dedicated to something beyond the normal work at hand. "Teams" go by other names as well, such as task forces, steering groups, and—shudder—committees. The bottom line is that they are all groups of people, with a "team leader," and usually focused on a specific task or mission. And they operate on a number of unwritten rules, which we will discuss, that are critical to successfully accomplishing the work of the team.

Some of these rules involve: believing in the basic premise of working through and with people; adopting the mindset that everyone needs to think of the whole team first, not just themselves; giving people the "big picture" about a project; dividing the work fairly; assigning all team members responsibility for tasks, which then makes them accountable; and learning to say "Thank you" to team members for doing a good job. Showing your thanks is, of course, not limited to recognizing team members, but needs to be an important thing for you to do in many instances when thanks is deserved.

> Working effectively and efficiently as a team is very important for a young engineer to learn.
>
> —Engineering manager, DMJM+Harris

Let's start with a few words about why working in teams can be difficult or downright uncomfortable for many engineers. One reason is that the teams you formed in engineering school—to work on senior design projects, professional society committees, or other efforts—were very different than the teams you'll find in the real world. Teams in engineering schools simply do not require that everyone be accountable for doing their work and contributing to the team effort. Sure, there were grades for the design projects, but the professor couldn't really tell who did the work and who did not. Some professors (including myself) occasionally resorted to asking each team member to grade the efforts of the other members. In the real world, teams are accountable for completing work and the team members are, by direct extension, accountable as well. Teams in engineering school do have value in enabling people to work together under the guidance of the professor, but their value is not directly transferable to the real-world teams in which you will be involved.

WORKING THROUGH AND WITH PEOPLE

The basic underlying concept of working in teams lies in the belief that people coming together can accomplish better things than by working alone, whether because of the opportunity to draw talent from various disciplines or other functions, or the notion of positive synergy that results from "two heads are better than one" and a "half-dozen or more heads . . ."—you know the rest. Given some of the difficulties of working effectively in teams (about which, more below), and because many young engineers are more comfortable working alone, working in teams requires a *belief* that you will accomplish something more than the sum of the parts, also known as "synergy." One simply needs to buy into the concept that you can accomplish more by working through and with people, as long as you respect everyone's situation as much as possible in this often informal and short-term setting.

Working Effectively through Teams Requirements

- Working through and with people
- Thinking "we," not "I" (There's no "I" in "TEAM")
- Getting a team to work well together by
  - Giving the team the "big picture"
  - Dividing work fairly—don't be unfair
  - Assigning responsibility to achieve accountability
  - Saying "Thank you" and giving special recognition if deserved

---

One simple reflection of this people-oriented view of teams is that you *must* think in terms of "We," not "I." A slogan you may have heard related to this: "There is no 'I' in T-E-A-M." You truly must set aside, or at least keep to yourself, your own personal agenda when you are working in teams. I feel it is helpful to think of sports teams, which can only be effective if each player knows their respective responsibilities and works together with others. This does not mean that all "teams" adhere to this, and some are more successful than others. There are no absolutes here, only people coming together to accomplish something that individuals or other structures could not.

GETTING A TEAM TO WORK WELL TOGETHER

On the surface of it, teams should be able to work together effectively if the people are reasonable, cooperative, and in the same organization. That may be true, but here are some things that I believe one must do to ensure that people work well together. They are mainly focused on what you should do as the team leader, but also include areas to assert yourself as an active participant of the team.

Give the Team the "Big Picture"

Everyone wants and needs a good understanding of the overall issues and reasons for the team, in the context of the organization's goals and objectives. The reverse of this, which often happens in teams, is that everyone is put on a "need to know" basis, that is, you only hear what someone thinks you need to know and nothing more. Also, since teams are often formed with people with different specialties, say, with you as the engineer, you might be only informed of technical issues, since it is assumed that you aren't interested in the marketing or the financial aspects. I believe this is wrong and leads to the frustration of intelligent people who don't want to be treated like children; they want to be aware of the big picture. (This often translates into frustration and complaints that their organization is "not communicating enough" because they have a perception that they don't know what's going on.)

It's actually easy to keep everyone informed of the status of the team effort and where it fits into the scheme of things. Of course, you must understand the importance of it first. Simply begin each meeting or status report with an overview of where things stand, including the simple repetition of what the team is all about. You cannot or should not assume that everyone is on "the same page"; in fact, there might be new team members, or someone not on the team and hearing about its status for the first time.

Here's something you can do if you are leading a team to give team members greater confidence and respect. Say at the first meeting that everyone is expected to be able to give a briefing on the team's efforts, not just their own part. Stress that the work of the team is a "shared responsibility," that we are all in this together, and that no one is smarter than anyone else. Explain why you have been given the task of leading the team. This approach does some wonderful things, most not very visible. It instills in each member greater self-esteem by showing them that the team leader views them with respect, and takes away a bit of their fear that they won't understand what's going on beyond their area of expertise. One of my most professionally satisfying experiences happened as a result of this sharing approach.

I led a selection panel to recommend an airport advertising contractor in a Request for Proposals process. This was an extremely competitive tender, with large amounts of revenues at stake, in a very complicated business activity. It did not appear that it was going to be fun working on this team. I was therefore very careful to keep the team aware of the big picture, and we all learned as much as we could about the "out of home" advertising industry, airport advertising in general, and the relevant activity at our airports, all presented in a "white paper." This served as the baseline for our evaluation of proposals that were submitted. In my signal to team members that each should be prepared to brief others on the project, I specifically took aside Pat, our financial expert, a young up-and-coming professional, and alerted him that at some point when our director asked me for an update, I was going to ask Pat to handle it. Pat didn't really think this would happen, until a day near the end of the effort when we were walking in the corridor and bumped into our director, Jerry.

Jerry immediately asked me what was happening with the selection, and I turned to Pat and said "Pat, why don't you brief Jerry?" Jerry looked at Pat and then at me, and then back at Pat, who calmly gave Jerry a status briefing. Jerry listened, then looked at me and asked a question. I again looked at Pat and asked him to respond. Pat answered Jerry's question. End of story. Although this may not be important in the cosmos, what happened was that Pat performed very well and was empowered to show what he knew, and I was proud that Pat took on the responsibility. I delegated some of my "power" as team leader to make someone else "look good." This is an important part of being effective as a team. In basketball, the star point guard is sometimes supposed to forego his or her shot to feed the team's top scorer, so that the team can be as successful as it can.

Divide Work Fairly. Don't be Unfair!

Perhaps no aspect of accomplishing work in a team is more important than allocating work tasks fairly. The converse is true—if work is divided *unfairly*, it will undermine the motivation and energy of the team members to make their best efforts. But, as said elsewhere in this book, this is easier said than done! Let's talk about what's "fair" and what's "unfair" for a moment.

> I need to work more with teamwork and working with others as I am a very individualized person.
> —Senior engineering student, Webb Institute

I believe it is very hard to be "fair" in the strictest sense, especially when assigning work in teams. Very few teams have work tasks so clearly set out in ways that one can allocate work evenly in quantity and amount of time required. Several cases that come to mind are when a finite number of activities have to be done, say, review and select from a certain number of proposals, where all team members need to read everything and perform similar reviews. Another example is when a team has to check references, and can evenly divide the (say) 10 people to be contacted among the five team members. In fact, the extent to which the work of the team *can* be divided equally will help ensure fairness and equal effort among the team. However, this cannot always be done with most team efforts in the complex real world.

More often, we're faced with simply trying not to be unfair in dividing the team's work. How does one measure "unfairness"? I believe that it is not a matter of measuring; rather, it will be apparent to the person or to the team if someone or several team members are being treated unfairly. This may relate to the fact that some may have inherently more work to do by the nature of the expertise they bring to the project. For example, financial members on a project may have to do more analysis than other members, and engineers may have to spend more effort on the technical aspects. The best way to deal with this is to be open about it. Let team members tell the team and the leader if there's a problem in work allocation. The team leader may be able to adjust deadlines and/or get more resources in specific areas. The important thing here is to build a feeling of trust among all the team members that is not undermined by any feeling of unfairness.

Assign Responsibility to Achieve Accountability

The natural extension of being fair in allocating tasks is assigning responsibility for specific tasks, which leads directly to accountability. This is crucial for a team to effectively accomplish its mission. Unlike in school projects, where there are fewer consequences if all the team members do not produce, in the real world it is critical, and often the cause of failures of teams.

The best way I've found to ensure that responsibility is clearly defined and assigned is for the team leader to think about and design the work plan of the team's effort to identify the specific tasks and products, and when they will be needed to complete the work. Many of these tasks and due dates may be obvious, and some may not be so obvious. This will lead to individuals being solicited to join the team, and then, importantly, agreeing that they will accept responsibility for completing the work task in the time specified. This is a "fitting process," in that the person may still have continuing responsibilities from their other job if this is not a full-time team, may feel like they need more time or help to meet the due date, and express other concerns about the scope of the task they're being asked to accomplish. To get their buy-in, they certainly should be able to have as much input on how and in what time frame their tasks are to be accomplished.

That being said, the team member's agreement that they will be responsible for specific tasks commits them professionally to do the work, and makes them accountable for their completion. "Accountability" is a very important principle in the real world—that one will accept praise for a job well done, or criticism or worse for poor or unfinished work. For most members of a team, this will be reflected in future compensation and performance ratings. In the reality of the team, a good way to keep everyone driving toward successful performance is to apply "peer pressure"—have all team members report their progress at every team meeting or status report. The team leader can use these reports to monitor progress, and get early warning if a team member is falling behind or is otherwise unable to accomplish the tasks on a timely basis, and then take corrective action.

> Do people (team members) know they are responsible for their work even though you are the head (CEO) of the team?
> —Engineering student, Binghamton University

Let me illustrate this with two examples. Early in my career, I chaired a professional society committee, which, by the way, is a wonderful opportunity for all engineers to get involved and grow professionally! Our committee was to investigate an issue and there were about 10 aspects to this issue. The committee consisted of about 20 members from all over the United States and several foreign countries, most of whom I had never met. I thought about how to get an effective response and input from the committee and do this in a reasonable amount of time, and sent a message to the committee in which I assigned two members to *independently* study and make comments on each of the 10 issues, by a specific date. I reasoned that if I got only one response on each issue by the specified time, I could knit together a report. If I got two responses on the same issue, I could meld them together and/or offer differing opinions. If I got no responses from either person on an issue, I would resolve to do the work.

This worked out very well. I received input from about two-thirds of the committee and was able to prepare a report, which was sent to all to review, on a timely basis. I believe that the reasons for this success included: an even and fair division of work, planning for some redundancy in the expectation that some would not respond, delegating responsibilities to the team but being prepared to "back stop" and fill in where required, and setting firm dates for action. (As a corollary, I sent a message to those committee members who had not responded at all, saying that since they were apparently too busy to respond, I would assume they no longer have time be on the committee unless I heard from them by a specific time. This is a way to keep committees and teams fresh and productive.)

Another example has to do with appealing to one's personality and professionalism to motivate team members to take on work when they might otherwise be disposed to be negative. It builds upon the discussion in the previous chapter on our personalities and how it affects our work. I was managing a voluntary team effort that was investigating ways to improve ground access to airports, and I learned that an airline was doing something very interesting from their Manhattan "city ticket office": passengers who were departing later in the day were allowed to check in their bags for departures from Kennedy Airport so they could spend more time in Manhattan before heading out to the airport, without the burden of their bags. At the time—even before the increased security in our post-9/11 era—we thought that this was not permissible. In fact, Cliff, the staff engineer who was involved in this area and a resource to our airport access team, said it was not possible. So I went to Cliff and said that though I understood his position, in fact the airline was doing this, and I was sure that they had received proper authority to do this. One of us should check this out, and it probably would be fun (that's the word I used) for him to call someone and then get back to the team. If he couldn't, then I'd have to do this at some point, because it was something we really needed to know.

I then paused to let Cliff think about this, and after a moment he said he probably could find out about this and write something up, but first he'd have to check with his supervisor to make sure he could spend time on this. I thanked Cliff very much, said it would be most helpful if he could report to us in about week, and went back to my office. Soon after, I got a phone call from Cliff's supervisor who blurted out: "Carl, how in heaven's name did you get Cliff to agree to do something extra! He always resists taking on extra work!" Cliff eventually produced a technical memo explaining how the airline was able to do the baggage check-in service, which proved valuable to our team's work, and I thanked him for his help. The importance of thanking people will be discussed next.

Say "Thank You" and Give Special Recognition When Deserved

One of the most important "props" I have in my business "toolbox" is a simple "Thank You" note—a note card that you can buy at any greeting card store in packets. It underscores to me the incredible and obvious value of thanking people for things; here, a job well done. To some readers, this will seem like "belaboring the obvious"

Figure 9.2   Thanking people on the team for a job well done is a wonderful gesture of appreciation. Keep "Thank You" notes handy for such occasions, and use them up!

and to others it may seem like being manipulative. I'll just say that we all appreciate our efforts being recognized by our managers, colleagues, family and friends, and this is not done as often as it should be.

Let me share with you the unusual way that I show a "Thank You" card as a prop during my seminar. During my discussion of "teamwork," I either put the card down on a surface as if lying on my desk, or hold it in the air that way, with the writing on the card facing upside down to the audience. The card I use happens to have large lettering, in capital letters: "THANK YOU." Only those closest to me are able to read the card. (It's the same card shown in Figure 9.2). So what's the point?

I then tell the audience that I've used this note card as a "prop" on my desk for many years, so that people who visit me can see "Thank You" cards prominently displayed on my desk, and know that I feel it's important to thank people. It's a reflection on how I do business, on my high regard for other people, and it says something about me, just like other pictures and mementos in my office.

But, of course, the importance of all this is to actively thank people when they deserve the recognition. I make it a practice to send at least one "thank you" note a week, and regularly browse greeting card stores looking for new designs. Of course, we now have the wonderful invention of electronic cards, or "e-cards," on the Internet, and I've shifted somewhat to using this medium because of the wide variety of sentiments available and the speed of delivery. But *a handwritten note is still most appreciated by the recipient.* (We could have discussed the importance of saying "thank you" in several places in this book, but it seemed most appropriate to discuss how it helps foster teamwork.)

Team members must be thanked and encouraged throughout their effort, even if their tasks simply involve doing what they're supposed to do. Some may argue that people shouldn't be thanked for just doing their jobs, but I disagree. The issue is to bestow the *appropriate* degree of thanks. I routinely end many e-mail messages with "Thanks for your help!" whether to professionals, support staff, or friends. And I mean it. When the "thanks" should be at a higher level, then send a "thank you" note, whether you take pen in hand, send an e-card, or pick up a phone and call. *You have no idea how good this can make someone feel,* or, maybe you do, if you've been on the receiving end.

I can tell you firsthand how good this feels. After giving a high-pressure briefing to the Port Authority Board of Commissioners about airport parking rates, I got back to my office and "vegged" from the stress of it all. About a half-hour later I got a phone call—"Carl, this is Steve Berger. [The Executive Director, whom I never had spoken with on the phone before.] Great job on the presentation to the Board! Thanks!" He then hung up. That was it, and the shot of adrenalin and good feeling was indescribable.

> Being a better communicator and more of a people person when it comes to interacting with the engineering team and client cannot be overemphasized in terms of success and job satisfaction.
> —Engineering manager, Port Authority of New York and New Jersey

Sometimes, efforts should go beyond giving thanks to someone when they truly deserve special recognition. And, as you will see, when people feel appreciated, this can then bring about greater accomplishment from a team effort. During the same airport access effort, we had a small core team of six people augmented by the voluntary efforts of about a dozen organizational units. Every month, we'd hold a "pizza lunch" at which reports on progress were informally presented. (These were held instead of formal, stodgy progress meetings.) The core team hit on the idea of bestowing an award—we named it the "Access Aces Award"—to someone outside the core team who had done something "above and beyond the call of duty." We had our graphics person, Gene, prepare some simple artwork on which we'd affix a small "thank you" message and "copy" the person's manager and department director. The granting of this award became the highlight of our lunchtime meetings, and an illustration of the value of this occurred a few years later. I happened to go to the office of George, and first noticed that there was almost nothing on his walls (a sign of something, to be sure) *except* that he placed his Access Aces Award prominently behind him and centered on his office wall. I didn't say anything, but marveled at how valuable this recognition had been to him, and how he wanted to show it to others.

Working in teams involves some additional insights and skills that may not be very apparent, but that can make the difference between poor results or excellent accomplishments. Hopefully, some of the above guidelines will be as helpful in your efforts as they have been in mine.

SUGGESTED READING

Bennis, Warren. *Organizing Genius: The Secrets of Creative Collaboration.* Reading, MA: Addison-Wesley, 1996.

Land, Peter A. *How to Build a Winning Team, And Have Fun Doing It!* Dallas, TX: Skyward Publications, 2002.

Toropov, Brandon. *The Art and Skill of Dealing With People: How to Make Human Motivation Work for You on the Job.* New York: MJF Books, 1997.

Vance, Mike, and Beacon, Diane. *Break Out of the Box.* Franklin Lakes, NJ: Career Press, 1996.

Woods, John A. *First Among Equals: How to Manage a Group of Professionals.* New York: Free Press, 2002.

CHAPTER 10

LEARNING TO NEGOTIATE

> Let us begin anew, remembering on both sides that civility is not a sign of weakness, that sincerity is always subject to proof. Let us never negotiate out of fear, but let us never fear to negotiate.
> —John F. Kennedy

I USED TO negotiate with each of my three children every year on their birthdays, but I didn't call it "negotiating," nor, to be sure, did I even know that I *was* negotiating. But now I see it for what it was. Each year I'd ask my children what chore(s) they would do to deserve a raise in their weekly allowance, now that they were getting older. This worked wonderfully. They had an incentive to tell me what things they would do for the increase, and I had a way to get some value for my money. Both parties were satisfied. To be sure, all the chores did not get done all the time, and, alas, I had to pay up or else! (By the way, you've gotten an extra benefit from this book: giving children an allowance at a very early age—about $1 a week when they first start school—and allowing them to increase it in this way gives them real experience in learning how to handle their own "spending" money.)

This is certainly not an example of how engineers negotiate— more on that shortly—but it is an example of a typical negotiation, nonetheless. For someone who didn't fully realize what "negotiating" really was until he was in his 40s, it feels strange for me now to be writing a chapter on "learning to negotiate"; recall the story about my personal epiphany at the negotiating course that appears in Chapter 1. So it would have been unbelievable to me during all those years I was *scared* of negotiating to think that one day I would explain to others the nature of negotiating and how to do it better. Well, here goes!

Stuff You Don't Learn in Engineering School. By Carl Selinger.
ISBN 0-471-65576-7 © 2004 the Institute of Electrical and Electronics Engineers, Inc.

Negotiating is simply the act of dealing or bargaining with one or more people to satisfy the needs of each party. Negotiations can range from informal conversations about where to have lunch, to deciding how to delegate work on a project, all the way up to intense wrangling over business contracts or labor disputes. You've been negotiating all your life, almost every single day, but often you don't realize it because no one had ever taught you how to do it—not in engineering school or in your work career. There is certainly an art to it, but also it's a skill you can develop, and there's nothing whatsoever to fear.

I believe that much of the psychological discomfort in negotiating comes from assuming that it is *always* a complicated and unpleasant task. My own fear of negotiation stemmed from my perception that it would inevitably be adversarial, such as bitter labor–management negotiations. This perception went against my people-oriented, congenial nature. In my case, my mother always told me that "If you can't say something nice, then don't say anything," which seemed to run contrary to what happened in negotiations. Understanding which situations require intense negotiating and which can be handled more informally is an important first step.

I don't recall many times when I engaged in overt negotiating during engineering school nor in the early part of my professional career. But, in retrospect, I now see that I did engage in negotiating on a number of occasions, now that I am a professor and on the other side of the proverbial table. Some examples of negotiating in college are: asking professors for more time to complete assignments; complaining that your test score wasn't right (or you didn't deserve to have so many points taken off); and, the clincher, arguing that you should have gotten a higher grade. Whatever the merits of any of these arguments, you were negotiating. Now, I won't say that negotiating in the real world is different from what it was in engineering school (but, of course, it is!), since the point here is to understand that you have always been negotiating, and now you should become comfortable in the skill of negotiating. In fact, the questions to your professor about due dates and the merits of your work are the kind of issues that do crop up in real-word negotiations.

CULTURAL DIFFERENCES IN NEGOTIATING

The purpose of negotiating something is to get to agree on terms that satisfy each party. But there are some *cultural* issues that affect this process right from the start. For example, unlike many other cultures around the world, Americans are not used to negotiating. For many Asians and Hispanics, bargaining over the price of things is imbued in their way of living. For those cultures, one *never* pays the asking or posted price—it would almost be insulting! You are *expected* to engage in a ritual of objecting to that price, even walking away, and negotiating down the price to something that both parties can honorably accept.

For most Americans, however, this is as alien as walking into a store, seeing a price on something, and saying to the retailer that the price is too much and you'll only pay a lower price. There have been some changes in these attitudes; for example, the growth of bargaining-oriented Internet activities like eBay, and name-your-own-price websites for air travel and hotels. But, except for buying a new or used car, bargaining for prices is not the rule in the United States.

Tips on Negotiating

- You negotiate every day!
- Identify your needs and the other party's needs.
- "Split the difference" is the American way, not necessarily the best way.
- Aim for a "win/win" to satisfy both parties.

Engineers engaged in multinational projects need to be more in tune with the cultural aspects of negotiating. But how does one deal with cultural differences that arise? One way is to try to follow the lead of others you are dealing with—try to do what they do ("When in Rome, do as the Romans do"). Also, have someone who is familiar with that culture join you in the negotiations or brief you on their negotiating customs. An interesting example of a cultural difference that would come up in negotiating with a Japanese person involves the meaning of the word "Yes." Yes, the meaning of the word "yes"! Let's say you indicate that the price for something will be $1,000. If the Japanese person responds "Yes" to you, the person doesn't necessarily *mean* "Yes, I *agree* with you. I will pay you $1,000." The person just as often simply means "Yes, I have *heard* you say that the price is $1,000" and, in fact, actually may not agree to pay $1,000. Negotiating with other cultures is often a very complex situation, so make sure you're well prepared.

NEGOTIATING PHILOSOPHY

Too often, negotiation is viewed by many as a game—a win/lose situation: one side *has* to win and, therefore, the other side *has* to lose. In other words, for one side to get what they want, the other side can't get what they want. The outcome of a win/lose negotiation may be just that: one side is happy and the other side is angry. Or it can lead to a hard-fought compromise in which the parties agree to "split the difference"—a normal outcome in the American culture—but they've found a middle ground that may satisfy neither party.

Isn't there a way to negotiate that leaves all parties satisfied? Yes, there is. It's called *"win/win" negotiation.* How do you achieve this? For relatively straightforward situations, like negotiating work hours or due dates, it's important to first have a clear understanding about everyone's goals: what *you* need, and what you think the other party needs. In more complex negotiations, the same rules generally apply, but with perhaps more issues and parameters to deal with, and often more opportunities to find an outcome that satisfies all.

A negotiation begins by first doing your homework. Start by writing down what you need (what you must get) and then what you want (what you would like to have). Say you're negotiating a work arrangement to telecommute because you need to be home several days a week (perhaps to take care of your children). So, your goals are: "I need to be at home two days a week, preferably on Mondays and Fridays, and I want the company to pay for high-speed Internet service at my home to help me do that." Then write down your fall-back position, which is what you'll settle for if you can't fully achieve your goals. Include on your list as many terms as you can. You can even include items that may be considered intangible; for example, that you'd be available to "attend" meetings or come into the office on a Monday or Friday under certain circumstances, or that you'd be willing to take on another added responsibility in order to make this telecommuting arrangement work.

How do you go about estimating the other party's needs or goals? This often is not too difficult since you usually know the other party and can intelligently guess at their needs. Try putting yourself in their shoes. You can also talk to people who might have more information or insight into the other party. And, of course, you can *ask* the other party before the negotiation begins, or at the start of the negotiation.

Figure 10.1 Understand each other's needs when negotiating things like due dates and tasks, and don't be afraid to discuss these things at meetings.

When the actual process of negotiation gets under way, you already know what your needs and wants are, and have "guesstimated" the other party's. The key at this point is to look toward a win/win outcome, whereby both parties achieve their respective needs and wants, rather than moving toward splitting the difference, or worse, adopting an immovable stance (like "take it or leave it").

Is it impossible to reach agreement? No. Even if both parties are widely apart on key issues, there are, in many cases, other issues on the table to allow an opportunity for trade-offs—"I'll give you this if you give me that." ("I'll agree to work from home only one day a week if you agree to pay for my childcare provider for that extra day.") This give-and-take is different from "splitting the difference" because you're not just compromising on one parameter that may not satisfy either party. The eventual outcome you are aiming for is that both parties reach a satisfactory and robust settlement that covers all the issues. There's definitely an "art" to negotiating to go along with the skill.

Here's another example from my own recent experience: negotiating with my editor over the due date for the draft manuscript for this very book. When we first discussed the book, I agreed to submit the first draft manuscript about six weeks after the contract was signed, so that the publisher could quickly send the draft out to reviewers. (Yes, this was a pretty ambitious deadline, but, hey, I had never written a book before!)

When it became apparent to me that I would have trouble meeting the deadline, since other unexpected business activities intruded on my writing time, I talked with the editor about pushing back the due date for the draft. I knew I needed more time to finish the manuscript, but I also knew that it was important to her that the manuscript be submitted on time to ensure a timely publication. We came up with a compromise during a brief negotiation in which we found ways for both of us to satisfy our needs by figuring out a different way to handle the review process: I would submit the first half of the draft manuscript on a certain date, so that reviewers could take a look at it, while I completed the rest of the manuscript for their later review. My editor and I, in other words, worked out a win/win agreement that satisfied both our needs, without having to argue simply about changing the due date for the entire submission.

Now let's identify a few typical negotiation scenarios that engineers face and suggest approaches on how to deal with them.

Due dates are a very frequent issue—when something you will do needs to be done, or when a "deliverable" is due from a contractor, and so forth. The negotiation should always be very specific on when something is due, so both parties are clear on this. Ways to deal with changes in due dates could involve adding more resources to finish sooner, agreeing to add more value if an extension is allowed, or figuring out a whole new schedule of deliverables.

Changes in the project scope can involve deciding to reduce the scope of tasks to finish earlier, adding more resources to accomplish a larger scope, or reorganizing the approach to the job.

Making an appointment or setting a meeting is, perhaps, easier, comes up often, and is indeed a negotiation. Try having both parties identify several possible dates during a certain time frame and then find two dates that work. Set the meeting for the earlier date and calendar the later date as a back-up if a problem arises with the first date.

Other areas in which engineers negotiate include how to respond to your manager when he/she makes a suggestion for you to consider, how to deal with change orders, how to handle conflicts between several projects you're working on, and, of course, many things involving money.

You now hopefully see some of the kinds of negotiating situations that crop up on a daily basis. So be ready to negotiate when a situation arises that requires you to resolve differences and to achieve your goals. Most importantly, don't shy away from this necessary skill—work at it and make it work for you to be more effective and successful in your career.

SUGGESTED READING

Cohen, Herb. *You Can Negotiate Anything.* New York: Bantam Books, 1982.

Fisher, Roger, Ury, William, and Patton, Bruce. *Getting to Yes: Negotiating Agreement Without Giving In.* New York: Penguin Books, 1991.

Nierenberg, Gerard I. *The Art of Negotiating.* New York: Barnes & Noble, 1995.

Stone, Douglas, Patton, Bruce, and Heen, Sheila. *Difficult Conversations: How to Discuss What Matters Most.* New York: Viking, 1999.

CHAPTER 11

BEING MORE CREATIVE

I need to be more creative. I tend to get stuck in the proverbial rut.

—Engineer at DMJM+Harris

I ONCE HEARD the inventor Paul MacCready give the commencement speech at The Cooper Union in which he discussed his invention of the human-powered flying machine. His "Gossamer Albatross," piloted by cyclist Brian Allen, actually crossed the English Channel in under three hours in 1979. So, if any person could be regarded as "creative," Paul MacCready fills the bill, and yet his talk was about how to be more creative, and he used himself as an example of the difficulty.

He told a story of how a youngster described to him a science project in which the challenge was how to float a straight pin on the surface of a glass of water. Since the idea was to allow the water's surface tension to keep the pin afloat, MacReady's thought process focused on how to gently lower the pin onto the water's surface so that it did not sink. He recounted to the audience a number of ways that he tried to do this, all unsuccessful. Finally, stumped, he asked the youngster how he did it. The answer: he froze the glass of water and then laid the pin on top of the hard surface so that it stayed afloat when the water melted. *Voila!* (That's French for "Yesssssss!") And you *can* try this experiment at home!

So, what is "creativity" and how can we become more creative in our lives? We start with a dichotomy: many correctly argue that engineering is a very creative profession, using principles of science to craft ingenious technological designs to solve problems. At the same time, there is a contradiction that engineers are not very cre-

Stuff You Don't Learn in Engineering School. By Carl Selinger.
ISBN 0-471-65576-7 © 2004 the Institute of Electrical and Electronics Engineers, Inc.

ative, that they are hostage to the state of the practice, and forced by codes or enduring traditions to do things the way they've always been done. Unfortunately, some would refer to a "creative engineer" as an oxymoron, and the cartoon engineer, Dilbert, can be cited as Exhibit A.

But I'm here to say that engineers *can* be more creative. The question at hand is *"how?"*

First, let's start with the ephemeral concept of "creativity." Dictionaries define creativity as the quality of making, inventing, or producing, rather than imitating, characterized by originality and imagination. (Notice that there was nothing in the definition about thinking inside or outside any "box," a faddish "buzzword" that I do not find very helpful.) But first a story about creativity in the workplace that might have easily appeared in a Dilbert cartoon.

A while back, in the middle of my career, the director of my department was very angry about his perceived lack of creativity on the part of our staff. Taking a typical corporate response, he formed a "Creativity Committee" to address this and come up with creative ideas, and I was put on this committee with a dozen others. Then, the director designated Roy (who probably would have been voted the most uncreative person in the department!) to chair the committee. At the first meeting, Roy announced to all that we were there to identify ways to be more creative, and asked if anyone had any ideas. No one said anything. I was the youngest member of the group, and so did not want to open my mouth, and was very uncomfortable with the whole business. I recall that the meeting lasting about an hour, accomplished nothing, and that, fittingly, never met again.

Well, I'm happy to tell you now that since that meeting long ago, I've learned that everyone is creative! Even engineers! The key is to first understand *why* engineers appear not to be creative, and then use some practical ways to unlock your creativity to improve the quality of your ideas and enhance your accomplishments.

One key to being more creative is to *look at things from different perspectives*. In his unsuccessful attempt to solve the pin-on-water problem, MacReady relied on his thinking process, perhaps constrained by some ingrained habits, not necessarily bad. In fact, each of us has formed our own approaches to these situations. Usually, we are guided by a number of set assumptions or parameters that we probably can't even articulate. These can include: being imbued with the current "paradigm"—the way it's done today—to address a situation; the culture of our organization or our society that guides or, really, constrains, our thought process; and, yes, our core engineering design approach, entailing specific step-by-step procedures that we've learned in engineering school.

This gets to the heart of why I believe engineers are not thought to be creative. Why is this? Let's give some simple examples. Aerospace engineers designing the next generation of aircraft cannot and should not "start from scratch." Most often, they need to build upon the latest state-of-the-art and try to improve incrementally on the performance, applying known principles of science and advances in technology. Engineers of all stripes use codes and standards that have meticulously evolved from new technologies, apply the safe state of practice, and learn from failure.

So where does "creativity" fit into this picture? Few of us are ever involved in conceptualizing new products or designs starting from the proverbial blank piece of paper. But every engineer's core mission is to try to improve the utility of things—to design something that will solve the problem with something better, faster, and/or cheaper. To instill creativity into this mix, the key is to un-

Figure 11.1 "What if . . .?" Coming up with creative ideas often requires concentrated thought and quiet contemplation.

lock new ideas with the use of *"What if"* or *"I wish"* statements. Make these phrases a routine part of your working vocabulary. When faced with a challenge to improve something or someone, write down a bunch of "What ifs" and see where your thinking process and imagination take you.

To be a little bit more concrete, but in the same vein, get more ideas by regularly talking to different people: others in your industry, people from different cultures, in different professions like law and business, and, yes, your customers or clients.

You never know where or when you'll get an interesting idea. I was once met at the Seattle–Tacoma airport by my cousin Howard, when "meeters and greeters" could still come right up to the gate to meet arriving travelers. As we walked toward baggage claim, Howard asked me what I was doing these days, which was airport consulting. When he heard this, he launched into telling me of an idea he had when traveling in Europe recently. It seems that there was going to be a long departure delay, and the airline staff told passengers to go shop or eat in the terminal. To ensure they were able to return at any time, they gave each passenger a "beeper" device that would page them when the flight was ready to board. Howard thought that was a cool idea, rather than having to wait at the gate for fear that the flight would board and leave before they could return to the gate. Why couldn't such an approach work at U.S. airports, Howard asked. I was really taken by the idea, and agreed to "talk it up" when I got back to work. When I did, by the way, I got no interest to try it, though many restaurants now routinely use these vibrating pagers to alert waiting customers that their table is ready.

> How do I come up with new things or look at things in new ways without "reinventing the wheel" or otherwise wasting time?
> —Engineering student at Tau Beta Pi seminar

Talking to all sorts of people and keeping your eyes open when you're traveling—and spotting interesting developments around the world—are rich sources of possible ways to do things better. Talk to visitors, colleagues, and people you meet about what you do, what problems they see, what is done in their field or country, and so on. Keep your mind open to the "What if we did it that way?" and you will find that some interesting and feasible concepts will bubble up to the surface. A new technology, or some incremental advance in the state-of-the-art, should prompt you to think "How does that change now enable me to solve this problem, or do this better, or seize an opportunity?"

How to Be More Creative

- *Everyone* is creative! Even engineers!
- Look at things from different perspectives.
- Talk to people from different cultures and professions, as well as your customers/clients.
- Unlock ideas with "What if " and "I wish" statements.
- Use Synectics and other creative processes.

Perhaps some lessons from my experiences as a "professional idea developer" may help you in being more creative in a practical sense. I spent many years managing "new business development" for the New York–New Jersey regional airports. My business card held my motto "Ideas are always welcomed!" and, indeed, many ideas to improve airport services found their way to my in-box for me to consider. (Holding yourself open to new ideas will also prompt people to feel comfortable when suggesting them to you.) To quickly assess each idea, I adapted the tools of the venture capitalist to answer things:

1. Can I understand the idea? Does it make sense? (If it doesn't, then stop right here and rethink the idea!)
2. Is there an identifiable market that needs or wants this product or service. Don't rely on a "Build it and they will come" mentality.
3. Do the people proposing the idea have a "track record" of demonstrated experience to accomplish the idea?

These are essential "success factors" needed to take ideas from the concept stage to the real world demand for practical application.

Let's talk a moment about how to encourage creativity, and, certainly, not stifle it or make it unwelcome. A normal reaction on the part of many when pitched a new idea is to think of all the reasons it will not work. Oftentimes, the smarter the people, the more reasons they will be able to offer (or conjure) up . . . whether based on legitimate issues, their comfort with the status quo, bureaucracy, the ever-popular "We tried that already and it didn't work," and the prideful, unsaid "not invented here" mentality. I can recall numerous times where even I was presented with an idea that seemed outlandish and impractical, and fought for the right words to say in response. So, here are two very powerful questions that I have used that will unlock the essence of the idea:

- *"What made you think of that?"* will get at the root of "how" the person came up with the idea and their thinking process, and/or perhaps lead you in a different direction or to other ideas.
- *"What are you trying to accomplish by doing that?"* will smoke out the goals of the idea, and the underlying assumptions being made, and, again, suggest different directions in which to probe.

There are also a variety of processes one can use to be more creative that probably have a certain appeal to the engineer who is comfortable with systematic ways to solve problems. These include various "brainstorming" efforts and other approaches that I will not cover here, with one exception. People who have been involved with brainstorming efforts generally do not come up with satisfying or practical ideas, otherwise it would be used more. People feel uncomfortable with the spontaneity; some people dominate with their ideas while others are reticent to jump in and perhaps get ridiculed for their ideas. It is also difficult to tabulate or work with the ponderous list of things that get disgorged. Oftentimes, there are no "next steps" after someone simply records the ideas. There is at least one better way.

What if we combined the brainstorming approach with a more

The most important character traits that I would like to see more of are ambition, enthusiasm, and proactive thinking. Engineers coming out of college seem to look to the experienced staff around them and seem to want to follow their example more than lead. I would like to see more creativity in their thinking and more expression of their own ideas in problem solving. I would like to see more enthusiasm in tackling and finishing each project assignment and being anxious to apply what they have learned to their next project. This relates to their own confidence in their interpersonal skills and in their own abilities.
—Engineering Manager, Port Authority of New York and New Jersey

disciplined harvesting of ideas addressed to a specific problem or issue that is offered by a client manager who has the authority to act on ideas that seem feasible? In other words, we need a process to transform a wild, free-flowing brainstorming session directly into specific actions. Dare I call it "structured brainstorming"? This process is called "Synectics" and has been around for some time.

Here's how a Synectics process works. A client manager brings a specific and serious problem or issue, which has previously defied many attempts at solution, to a group of some half-dozen or more people not involved in the specific issue who are bright and are not afraid to speak their minds (yes, people who are viewed as "creative" types). After the client manager explains the problem/opportunity, the group silently takes a period of time, say about 15 minutes, to write down a list of things that come to mind in the form of "I wish" statements. (They do *not* speak out loud at this point, as would happen at a typical brainstorming session.) These "I wish" statements can be practical thoughts or, more typically, wild-eyed concepts that come to mind. After all, there would be no Synectics session if the obvious practical thoughts had worked. At the risk of using one simplistic example to illustrate this, imagine that at a session held to identify ways to make public telephones vandal proof, one person wrote down "I wish the payphones were as invulnerable as a Sherman tank." Yes, that would probably solve the problem, but be a bit costly.

At the end of the silent period, each person has written down many "I wish" notions, probably at least one or two dozen. Since these are written down, they actually become the fodder for easily recording them at the end of the session for future contemplation. Then the fun begins. Each person starts reading off what they wrote and someone records them on flip charts as "I wish." Importantly, *no one is allowed to criticize any idea at this point*! However, if the idea is really far out, participants can ask what made them think of the idea, since that could cause others to spontaneously add onto the idea or suggest an additional thought, much like in a regular brainstorming session. This process allows everyone to get their ideas out on the table, and then provokes further, often very animated, discussion and creative thoughts.

After some period of everyone disgorging their ideas (and disgorging is a good way to describe what happens!), the process moves on to the next, more disciplined step: to sort out and combine like ideas into main categories. The client manager begins to interact with the group at this point, reacting to areas of fresh thinking that have serious potential, and these areas will be the products of this creative effort. The next steps depend directly on the client manager who, you will remember, has the authority—indeed, the *responsibility*—to implement feasible suggestions.

Can this type of process work with engineers to improve creativity in addressing engineering issues? The answer is yes, and here are two examples. When I was looking for new ideas for airport improvements and services, I asked the engineering department manager if it would be okay for me to facilitate two separate 90-minute meetings with their electrical engineers and their mechanical engineers, to identify new ideas. I intended to use the Synectics process with each group, and made a big point to insist that the meeting would take no longer than the 90 minutes so as not to take too much time away from their work. In each meeting, about 20 engineers arrived at the conference room, most having no idea why they were sent. (As mentioned in Chapter 7, they were just "told to go" to the meeting.) I told the groups of engineers that we were going to run a kind of brainstorming session to identify new technology ideas to implement at our airports, and immediately sensed great discomfort from many around the table—an unwillingness to sit through such a session, with some eyes rolling, some surely thinking that this would be a big waste of their time. I predicted that they were probably going to be uncomfortable for the first half hour or so, and then they were going to all start getting into the surge of creativity, and after the 90 minutes, at 11:30 AM, they would not want to stop; I would have to say that they would have to go back to work since I didn't want to get into trouble with the engineering manager.

This prediction came true on these two separate occasions. The individual engineers silently wrote down their "I wish" statements, then all started reporting aloud to the group, with my assertive facilitation to avoid criticism of ideas and keep things on track. Each session yielded about 30–40 separate ideas that were easily harvested from their written notes and refined on some flip charts. And I did have to force enthusiastic stragglers to leave the conference room at 11:30 AM. Several insisted on staying and discussing some of the ideas! (As a sidebar, part of this process can be used when a meeting gets stuck on something or is unable to figure out what to do. Ask people to take a moment of "quiet time" to write down some thoughts, and then have each say aloud what they wrote down.)

So, can engineers become more creative? Can we even refer to "creative engineers" without calling it an oxymoron? Yes, if engineers are open to new thinking and are willing to pursue ways to come up with fresh ideas to improve the products they design and build.

SUGGESTED READING

Csikszentmihalyi, Mihaly. *Creativity: Flow and the Psychology of Discovery and Invention.* New York: HarperCollins, 1996.

Drucker, Peter F. *Innovation and Entrepreneurship: Practice and Principles.* New York: Harper & Row, 1985.

Gelb, Michael J. *How to Think Like Leonardo da Vinci: Seven Steps to Genius Every Day.* New York: Dell, 1998.

Kriegel, Robert J., and Patler, Louis. *If it ain't broke . . . BREAK IT! And Other Unconventional Wisdom for a Changing Business World.* New York: Warner Books, 1991.

Nierenberg, Gerard I. *The Art of Creative Thinking.* New York: Barnes & Noble, 1982.

CHAPTER 12

ETHICS IN THE WORKPLACE

Responsibility, integrity, and the like are important. Although we often fail in these departments, we must strive to practice them. Practice makes the master. We each are responsible for being the best person and professional we can be.

—Engineering manager, TransCore

SOONER OR LATER, we all face a situation that doesn't feel right. It could come from a phone call, or an e-mail, or a personal conversation. It may go like this: you are told to handle a situation in a certain way, with words like "I want you to do it this way," or "Don't worry about that," or "Don't tell anyone about this." Immediately, something goes off in your head and, especially, in your stomach, telling you that this is wrong. What's going on? And what do you do about it?

At times like this, often when you least expect it, your body and mind are telling you that this is an issue of ethics—that there may be a conflict with your personal ethics, your values, your concept of what is "right" or "wrong," of what is "good" or "bad." This can definitely be a time of great consternation for every person, and for engineers it often involves issues of safety, liability, fairness, and dollars.

Today, with accounting and other scandals undermining our trust in major corporations, being an ethical professional is assuming growing importance in the midst of complicated issues. These issues include trade-offs in numerous areas of safety, the environment, and legal and financial issues in our real world. With our technologies becoming more intrusive and pervasive, concern for personal and corporate ethics remain very important issues in today's global society.

Figure 12.1 Getting a difficult call. What do you say? What do you do?

The days when an engineer's only ethical commitment was loyalty to his or her employer have long passed. The expansiveness of technology is such that now, more than ever, society is holding engineering professions accountable for decisions that affect a full range of daily life activities. Engineers now are responsible for saying: "Can we do it? Should we do it? If we do it, can we control it? Are we willing to be accountable for it?"

—Murdough Center for Engineering Professionalism,
Texas Tech University

So what does an engineer need to know about being an ethical professional? It helps to first understand what "ethics" means. Dictionaries refer to it as a body of moral principles or values, dealing with right and wrong and the morality of motives and ends. As we grow up, we tend to learn our values and moral principles from our parents, our teachers, our friends, and from observing issues and behavior in our society. In engineering school, we are subjected to a very rigorous education and are told the importance of doing our own work, not cheating, and performing to the best of our ability. But does this conform with the reality of the real world of engineering?

What kind of engineering ethics issues arise most often? Certainly, environmental and safety issues are paramount: accidents that kill or injure may be caused by knowingly faulty design, construction, or operation; not questioning that what has always been done is what one ought to do; being fair without discriminating; as an expert witness, giving the same testimony in a court proceeding, whether for the plaintiff or the defendant; where there is a question of legality, just because something is legal, is it ethical to do it? Other areas include acknowledging mistakes, conflicts of interest, disagreeing with your employer, and being honest and truthful.

We often read about major corporations being cheated by executives, students plagiarizing term papers from the Internet, elected officials behaving badly, and a myriad of "whistleblowers" calling attention to alleged wrongdoing in their organizations. This is very disturbing to the mostly idealistic young professional. After all, we come out of engineering school believing that we are ethical persons, and, fortunately, that seems to be the case, as we'll see next.

Let's start with a reassuring note. From a continuing, informal survey of engineering managers, managers were asked how important "ethics" are for their young engineers. In the view of these managers, ethics garners a high rating, around 8–9 on a scale of 1–10, with 10 being the highest. Then, when asked "What do you perceive as the ability of young engineers?" in certain areas, the managers' ratings of the ethics of young engineers are again very high—between 8 and 9. (More on this Engineering Managers Survey in the Appendix.) What I conclude from this, and my general perception, is that ethics is an important attribute for young engineers, and that they have demonstrated to the managers surveyed strong ethical values. So, it appears that engineers start off their careers on the right path. However, as they travel the road of their careers, they begin to come into more contact with ethical issues in their work and in their personal lives that test their values and choices of right versus wrong.

"DOESN'T FEEL RIGHT" SIGNALS AN ETHICS ISSUE

Fortunately, there does seem to be an innate way to determine if a situation involves a question of ethics. Something doesn't "feel right" in our stomachs because our brain is signaling that the situation conflicts with something in our values, our sense of right and wrong. In short, we may be confronting a situation in which it is not clear how to proceed. This may involve a seemingly innocuous direction, such as "You don't need to say that" in a situation where you feel in your gut that you do need to call attention to the point. It may be a question of safety, financial performance, or technical specifications. And it may be a very legitimate issue having nothing to do with ethical concerns. But you are young, and have little perspective and experience to judge the situation accurately. It is important to listen to your body and your gut, though, and raise your awareness in these situations and try to ask someone for guidance or clarification.

UNETHICAL IS NOT THE SAME AS ILLEGAL

It must be noted that something that is judged to be "unethical" does not necessarily mean that it's "illegal." This is not to split hairs or have us all run to law books, but an important distinction. There are laws—specific acts of legislation—that define what is against the law (though often not so clearly and that's why there are courts) and the society's penalty for committing the offense. Ethical issues, as opposed to legal issues, range from very minor social interactions up to major transgressions of behavior in the workplace or society, but fall short of criminal activity. The point here is that one must have some perspective when confronting an ethical issue, and assess it as best you can in terms of the choices you have.

When It's an Issue of "Ethics" . . .

- Something that doesn't "feel right" may be unethical.
- Unethical is not the same as illegal.
- Values vary greatly by culture.

ETHICS VARIES GREATLY BY CULTURE

Ethical issues can also vary greatly by culture, nationality, and religious beliefs. In our world, different societies can have widely varying value systems. In some cultures, for example, bribery is an accepted part of doing business, but this is abhorrent (and illegal) in many Western societies. This varying sense of what is right and wrong can be very difficult for all of us to grasp, and, when coupled with basic human behavior, adds to engineers' difficulties because they are trained to seek the "right" technological answers to solve problems at the lowest cost. Add to this the growing globalization of business and technology, and we have an increasingly complex mixture of ethical challenges.

SO HOW DO YOU HANDLE ETHICAL ISSUES?

Perhaps the most important thing that engineers can do when facing an ethical issue is to speak up. That is not to say that you should call your local newspaper at the first hint of something you don't agree with. It is to say that you should not hesitate to raise your concerns to the person who is broaching the issue, or to your manager, or to colleagues to get their reactions and advice. Getting more information on the situation and testing your own thinking on how to look at both sides of the issue (the ethical tradeoffs) can be very helpful and lessen your inherent stress and discomfort. It is wise to follow with your instincts to a certain degree, because you have "stamped" within yourself your value system of right and wrong, and this is usually a good barometer to follow. But the nature of many ethical issues is that there *is* no clear right or wrong, so you may be forced to accept a compromise or a point of view that may trouble you, but only after airing out your concerns.

Dealing with Ethical Issues

- Speak up and follow your instincts.
- Talk to peers and mentors.
- Read ethics "case studies" and columns in professional-society and other magazines.

READ AND TALK ABOUT ETHICS

I apologize if I have been a bit vague on specific examples and guidelines to be an ethical professional. That is, I suppose, the nature of ethics. Fortunately, there are many ways to get more comfortable dealing with ethical issues. Reading case studies is an especially helpful way to understand the varied type of issues, and there are more venues for writing on ethical issues, fueled by the growing interest in ethics within the engineering profession and society in general. Tau Beta Pi, the engineering honor society, deals frequently with ethical issues in its publications, as do many engineering societies. Columns on ethics appear regularly in the popular press, including "The Ethicist" in the Sunday Magazine of *The New York Times.*

Although reading about ethical situations will be helpful, be cautious because the generic ethics issues presented in case studies, or even the specific ethics issues discussed in the media, may not deal exactly with an ethical issue that you personally encounter in the real world. The situation you face is invariably unique, and you need to apply your value system, your experience, and, yes, your common sense to deal effectively with the situation.

> You usually have to give up something of value to get something of value, and with ethics, there frequently is no absolutely right answer, just a personal best answer, and it all comes down to you.
> —Murdough Center for Engineering Professionalism,
> Texas Tech University

So you will come to realize that there are no clear guidelines to what is "right" or "wrong," that sometimes they are a "moving target" in our always-changing societies, and that they are very much in the eye and the heart of the beholder. The important task is to be able to quickly recognize ethical situations when they first arise, and then assertively deal with them in the most comfortable way that you can. If you do this, you will be a better engineer, and also sleep better at night knowing you've done the best you can.

SUGGESTED READING

Cohen, Randy. *The Good, The Bad and The Difference.* New York: Doubleday, 2002.

Halberstam, Joshua. *Everyday Ethics: Inspired Solutions to Real-Life Dilemmas.* New York: Viking Penguin, 1993.

Harr, Jonathan. *A Civil Action.* New York: Vintage, 1996.

Vann, W. Pennington (compiler). *Engineering Ethics References.* Texas Tech University: Murdough Center For Engineering Professionalism, 2001. National Institute for Engineering Ethics (NIEE), website: http://www.niee.org/biblio-1.htm.

CHAPTER 13

DEVELOPING LEADERSHIP SKILLS

I need to develop my leadership skills . . . be willing to take on more serious responsibility without self-doubt.
> —Engineer at a major Midwestern engineering company

I DECIDED IN my thirties that I wanted to learn more about leadership, so I had my company library do a search on books and articles in the literature. They gave me a printout with about twenty citations that the library contained, with a short description of each. My initial reaction was this was a very impressive list, and I began to think about which books I would read. Then I noticed that each citation gave the number of pages. Perhaps it was because of my quantitative mind—the engineer in me!—that I added up the total number of pages in all the publications. The total was over 1100 pages! I suddenly had a different reaction: I didn't have that much time to read all this, and how much would I really learn about leadership after plowing through 1100 pages?

While many of the various nontechnical skills we've discussed have specific techniques you can use to do them more effectively, leadership skills present some less specific but basic concepts that will be important to your work and to your personal life—what has been called by others "the art of leading and the discipline of managing." Many young engineers start their first jobs in a technical position and gradually move toward becoming "emerging project managers"; they begin to manage projects and people, whether they pursue a "technical track" or a "management track" in their careers. You must understand, however, that your ability to lead people and to manage activities will be important to your success, whatever

Stuff You Don't Learn in Engineering School. By Carl Selinger.
ISBN 0-471-65576-7 © 2004 the Institute of Electrical and Electronics Engineers, Inc.

your intended career path. And, like other skills we have discussed, I believe that they can be learned.

IS THERE SUCH A THING AS A "BORN LEADER"?

Let's start with your ability to become a leader at some point. I don't think it is helpful, or correct, to think that some people are born with "it"—so-called "born leaders," having the charisma or other attributes ("the vision thing") that enable them to lead people—and that you as an engineer cannot aspire to be a leader, per se. Perhaps there are some people who have better innate abilities that make people want to follow them, but I think it is more helpful for all of us to realize that *we each can develop and learn skills that will enable us to effectively lead people* in our workplaces, in professional societies, and in our personal activities, should we choose to do so. I believe the opportunities abound for you to develop your leadership skills.

First, though, who comes to mind when you think of "leaders"? The list would probably include Presidents, elected officials, the chief engineer of your organization, your manager, religious leaders, community leaders, scout leaders. In fact, they are all "leaders" in that, by definition, people follow them. To the extent that they are *effective* leaders, then their followers will strive to implement their policies and programs.

I began to think more about the leadership abilities of people I came in contact with, as well as leaders like Presidents and military generals, and began to observe patterns and contradictions. Without getting into issues in this provocative subject that go way beyond the scope of this book, one could assert that President Bill Clinton was a very effective leader whose character was suspect, a seeming contradiction in terms. And a classic example of leadership is General Ulysses S. Grant, who was recognized as having superb leadership abilities during the Civil War, but exhibited poor political leadership in his scandal-ridden administrations when he became President. Same man, different situations. So is leadership situational? We won't resolve this here, but it shows that leadership is a very ephemeral concept, and it is in the eyes of the followers to determine whether to follow someone's directives with their whole heart and energy. A tall order.

Figure 13.1 Leaders have followers, and briefing everyone on your goals and plans is critical to getting everyone to buy in.

BUILD LEADERSHIP SKILLS BY TAKING RESPONSIBILITY

I believe that the key to becoming an effective leader is to build "leadership skills," mainly by taking responsibility in your life. And I believe these "leadership skills" are none other than the collection of nontechnical skills covered in this book, with some perhaps being more important than others to help you lead. The most important skills that support effective leadership must include writing, speaking, listening, decision making, and setting priorities.

You must be able to write your ideas clearly so others can understand and engage. You must be able to speak to groups or in one-on-one situations to inform, persuade, and deal with conflict. Leaders need to be able to listen effectively so they can understand the issues or concerns of people and they must have the "radar" to pick up on subtle issues. The ability to set priorities can focus the followers on the most important issues that need to be tackled, instead of wasting their efforts in less-important activities, since there are always limited resources. But nothing is more important, in my opinion, than decision making, since leaders are expected to be decisive.

> ### Leadership and Management Skills
>
> - You're not born with "it."
> - Build leadership skills by taking responsibility.
> - Get involved! Step up! Ask, "Can I handle this?"
> - Managers allocate resources to get things done.
> - Efficiency is the result of doing the thing right.
> - Effectiveness is the result of doing the right thing, which is more important.
> - Manage your time, but be wary of "time management."

So how does one develop leadership skills? I believe it is crucial to *get involved* and *take responsibility*! In the context of your work and personal life, I believe you need to participate in as many work and professional activities as you can, to develop better understanding of different activities and get the opportunity to work with different people. This includes serving on committees, taking on increased job responsibilities, performing voluntary tasks, and joining professional societies, and extracurricular clubs and groups. Early in my career, I took advantage of an opportunity to participate in a professional society committee at the local level, and got to meet other professionals that I normally would not have come in contact with. I ascended to chair that committee within several years, and it proved to be a springboard to other national professional society activities and a wonderful source of networking with other professionals.

At work, take every opportunity to ask your manager if there is something else you can handle. This is not meant to attract attention to your ambition, nor does it mean that you don't have enough to keep you occupied (which might be the case), but signals that you are trying to improve yourself and help your organization at the same time, and that you are not afraid of accepting more responsibility. You may have already done these things in engineering school, and know that people view these activities as measures that you are "well rounded" and that you are better able to work with others.

Perhaps you have a perception that leaders are, or should be, perfect and infallible. You place them on a pedestal, think they know everything and are never uncertain, and so forth, and that this inhibits you from moving toward positions of leadership. This perception is not correct nor is it helpful. Most leaders have the same concerns and uncertainties that we all have; you just don't see it often, if ever. I remember having a big problem with a project that forced us to ask for guidance from the Executive Director of my organization. I expected him, as a leader and with all his smarts and experience, to confidently give us the proper guidance to solve the problem. Instead, I was startled at how he openly struggled with us to determine a course of action—leaning back in his chair at one point, putting his hands on his head, closing his eyes, and saying aloud "What the heck do we do here?" (he actually used more colorful language!). It was a "learning experience" for me and burst my bubble that leaders know everything; but effective leaders will be forthright and want to try to do the right thing.

A common question is, "What's the difference between a manager and a leader?" This is a good bridge into our discussion of what "managing" is and what managers do.

MANAGERS ALLOCATE RESOURCES TO GET THINGS DONE

Managers are people who are made responsible for allocating resources—people and dollars—to get specific things done. Whether it's your current manager, a project manager, the manager of a sports team, or the President of the United States, the manager is a person who needs to accomplish defined tasks with the available resources and time constraints. That's what managing boils down to. Although there are many books out there on "managing" that may be helpful for you to read (some are offered at the end of the chapter), most of the support skills you will need are contained in this book.

> I'm most concerned about being able to recognize, adapt to, and overcome changing times and challenges.
> —Engineer at a major Midwestern engineering company

So, the difference between a manager and a leader, in my opinion, involves the difference between the *art of leadership* and the *discipline of managing*. (It's not my thought, but I agree with it.) Leaders are often managers, but they don't have to be; managers are not always leaders, despite our wish that they should be. Leaders mainly need to articulate a vision or goal of where things should be heading, whereas managers are tasked with trying to accomplish specific objectives with finite resources. You can be a leader without being a manager—in your activities at work, in professional societies, or in your personal life—in the areas of responsibility that you take on, even in your own personal development, since, as you'll recall, you are the Chief Executive Officer of you! (And don't you forget it!)

EFFICIENCY VERSUS EFFECTIVENESS

Let's change gears now and talk about one more concept that helps us be better managers: "efficiency versus effectiveness." These are two parameters that sound similar, but are really quite different, and both are important to understand in sorting out your life in the real world. "Efficiency" and "effectiveness" are often married in management jargon with a "versus" to reflect their differences. Simply stated, efficiency is doing the thing right, whereas effectiveness is doing the right thing.

Efficiency versus Effectiveness

- Efficiency is doing the thing right.
- Effectiveness is doing the right thing.

OK, that sounds catchy, but what does it really mean? With so many things on our plate today, we do need to be efficient, which means learning how to handle things well in a minimum amount of time. This could involve returning a batch of phone calls, trying to handle each e-mail or piece of paper only once, and handling repetitive tasks like preparing periodic reports in a minimum amount of time. Doing things more efficiently will get them off your plate quicker and save more time for other important things. Keep asking yourself: "Am I doing this efficiently?" or "How can I do this better/faster?" A tip here is to ask others—your manager, your peers—how they handle that type of task.

Early in my career, a more senior gentleman at our company told me: "This company focuses too much on doing things right and not on doing the right things." I [now] understand better what he meant. The optimum is to do the right things the right way.
—Chemical engineer at AIChE seminar

Effectiveness harks back to judging the *importance* of the tasks before you, and upon which your performance will be evaluated. Doing the right thing means that you will accomplish more important things in a given amount of time and at a given cost, and it is very much a product of your judgment. It does not measure how fast you do things, but rather whether or not you do the right things to achieve your employer's goals as well as those in your own career and personal life.

TIME MANAGEMENT

Before we leave the subject of developing leadership and managing skills, a brief word about "time management." I prefer to call this *"managing your time"* as we all have the same allotment of time in this world, and it is what we do with it that yields maximum value in our lives that is important. It is simply not a matter of optimally allocating minutes and hours. To be sure, there are numerous courses and books on time management, most having to do with recording your current minutes and hours over a week devoted to this task or that. I once took such a course, and found it very unhelpful and misguided, despite its obvious appeal to engineers who seek a quantitative way of organizing their lives to have more time to do things. My experience was that it was impractical to keep such meaningful records, and it did not apply to the judgments you need to make to manage your life, not just your time. For example, you probably need to learn how to say "No" to increased work you are asked to do, especially if it's lower priority compared with other things on your plate. How do you assertively and tactfully say "No"? There are several ways: "I would love to take that on, but right now I have too much on my plate." "I can't get to that now. Can it wait for a few weeks?" Or, "Well, OK, but if I take that on I would have to drop this other task."

> The challenge: How to manage time. Time to work. Time to relax.
> —Junior engineer at ASME seminar

To sum up, I firmly believe that the essence of developing your leadership and managerial abilities is to master the various nontechnical skills covered in this book, and then take increasing responsibilities in your work and personal lives to which you can apply these principles.

SUGGESTED READING

Covey, Stephen R. *The Seven Habits of Highly Effective People: Powerful Lessons in Personal Change.* New York: Fireside Books, 1989.

Crocker, H. W. III. *Robert E. Lee on Leadership: Executive Lessons in Character, Courage, and Vision.* Rocklin, CA: Prima Publishing, 1999.

DePree, Max. *Leadership Is an Art.* New York: Doubleday, 1987.

Depree, Max. *Leadership Jazz.* New York: Currency Doubleday, 1992.

Kaltman, Al. *Cigars, Whiskey and Winning: Leadership Lessons from Ulysses S. Grant.* Upper Saddle River, NJ: Prentice-Hall, 1998.

Kaltman, Al. *The Genius of Robert E. Lee: Leadership Lessons for the Outgunned, Outnumbered, and Underfinanced.* Upper Saddle River, NJ: Prentice Hall Press, 2000.

Penzias, Arno. *Ideas and Information: Managing in a High-Tech World.* New York: W.W. Norton, 1989.

Peters, Thomas J., and Waterman, Jr. Robert H. *In Search of Excellence: Lessons from America's Best-Run Companies.* New York: Warner Books, 1982.

Phillips, Donald T. *Lincoln on Leadership: Executive Strategies for Tough Times.* New York: Warner Books, 1992.

Safire, William, and Safir, Leonard. *Leadership: A Collection of Hundreds of Great and Inspirational Quotations.* New York: Fireside Books, 1991.

CHAPTER 14

ADAPTING TO THE WORKPLACE

> As a new and young engineer, I think the hardest part is learning how to adapt to a world that is predominantly much older than you.
>
> —Chemical engineer at AIChE seminar

I DISCOVERED THE importance of "turf" early in my career. No, we were not planting grass on the lawn outside. We were dealing with the very human-animal need for territory inside the office. I found that the workplace was, indeed, a very different place than school or home.

We were rearranging the office "partitions" in those days, much like rearranging cubicles today. Our group had reviewed the plans for who was going to move where and how much space we each would have—actually an increase in space for all. Everybody approved the plans and over the weekend the workers rearranged the partitions. Monday morning arrived and our group ambled in to start the work week and set up our new offices. Then one of our group arrived, took a look at the layout and location of *his* office, and suddenly became very agitated. At first my manager thought he was kidding, and then began to see how really upset he was. It all had to do with his (incorrect) perception that he had less space. The point is that this was rather trivial to the rest of us, but it was incredibly stressful to that person.

THE WORKPLACE—YOUR HOME AWAY FROM HOME

It's not "home," but you'll live there for a good part of your waking life. It's where you work—the office, in the field, wherever—the

Stuff You Don't Learn in Engineering School. By Carl Selinger.
ISBN 0-471-65576-7 © 2004 the Institute of Electrical and Electronics Engineers, Inc.

place you come to early in the morning and leave from late in the day. It's not like school either, and there are a number of physical and societal things that you'll need to understand to be more effective and happier in your work life. What kinds of things? Professors have been replaced by "managers," it's hard to know what's right to wear, there are all sorts of subtle and not-so-subtle relationships you need to deal with, and so on. We will try to explain some of these issues that the engineer faces in today's workplace, and suggest ways to deal with them effectively.

TURF

Perhaps no workplace issue is as subtle as something called "turf," a reflection of the primordial territoriality of our animal nature. This comes up in two general areas: the boundaries of your workspace, as we just saw, and a different form—the scope of your work responsibilities. Let's talk first about your workspace, which can be a real office, the proverbial cubicle, or increasingly, open areas. In each case, there is a mostly unseen "territory" that you have—almost like an invisible wall—beyond which you may infringe on someone else's "space." This can be very uncomfortable for either party, as it conjures up our animal instincts to protect our defined territory.

I won't dwell on dogs and fire hydrants here, but there is ample support for how our animal natures affect our real-life behavior. For an example, read *The Naked Ape* by Desmond Morris, an account of the traits and habits of humans as animals. And, more relevant to the importance of defining boundaries, poet Robert Frost famously wrote, "Good fences make good neighbors."

> I'm concerned about office politics. What do you do when something "serious" arises and it seems as though nothing can be done or said to help the situation?
>
> —Engineer at Tau Beta Pi seminar

A more serious issue of "turf" arises when one person "invades"—yes, that's a good word to describe it—the responsibility of another person or group. This engenders a response like: "Hey, that's *my* area of responsibility!" Translation: "You're *not* supposed to be doing that work, that's my job, so *STOP IT!*" How can this happen, and why do people react so vehemently to this type of "turf" issue? Many times, the transgression is quite innocent. Perhaps you were not aware that someone else was supposed to do that work. This can usually be remedied by a quick, soft apology: "Oh, I'm sorry I didn't know that you were responsible for that."

A Guide to the Workplace

- It's not home, but you'll "live" there most of your working life.
- Turf, copy lines, your office, the tyranny of the inbox, etc.!

This suggests a short digression. A very useful management dictum is that "It is easier to ask for forgiveness than to ask for permission." This means that is easier to apologize for doing something if it becomes an issue, rather than having to go through the process of getting approval and risk being told "No." Of course, this approach should only be used very selectively, with a good understanding of any potential consequences.

Other times, turf issues may reflect a real conflict in which two units or people *do* have overlapping or unclear responsibilities, in which case it has to be resolved at an appropriate level of management. But the "why" of turf (battles) is very important to understand: you are dealing with someone's basic job. That's why people regard this very seriously, especially if someone is *intentionally* stepping into someone else's turf, to try to do the job better, or to increase one's power in an organization.

"COPY LINES," ALSO KNOWN AS "CCs"

Have you ever noticed or wondered about the often long list of names of people who are "copied" on memos or e-mails? Who *are* all those people, and *why* do they need to get a copy of this message? Welcome to one of the rituals of the workplace, which changes from organization to organization, and is a subtle barometer of the culture. (We'll get to "culture" shortly!) First, let's describe what "copying" is. Simply speaking, it's giving a copy (print or electronic) of the message being sent to someone to others inside and outside the organization, each of whom "needs" to see a copy to be informed of the message. The nuance here is that some/many people want to be "copied" so they can be seen as powerful or influential, or simply be kept "in the loop" of knowing what's going on. In some organizations, this causes "copy lines" to be three or four (or more) lines of people who get the message. In this electronic age, this can be done instantly with e-mail, though trees still fall in the forests to support the printouts of unneeded copies of messages. (Mercifully for support staff, the old days are gone of the arduous task of making Xerox copies and then mailing them internally and externally to dozens of people copied.)

Now, if "CCing" or "copying someone" weren't enough to absorb for the engineer, we ought to mention "BCCing" or "blind carbon copying." (The "carbon" has to do with times not very long ago when copies were actually made with a sheet of carbon paper, much the same way that my fax copier still uses a cartridge to make imprints.) Whereas everyone involved sees the "CCs," a blind carbon copy is when you want someone else to get the message without everyone else knowing it. One example should suffice: you reply to a message from a customer and CC everyone who ought to see the message, and BCC your manager, say, so that she/he sees what you're doing. CCing and BCCing are very easy to do with e-mail these days.

To illustrate how important CCs are to understanding the organization, here is an experience one of my former students had in her first job with a municipal water resources department. She often had time on her hands, noticed all the CCs on memos she would read, and asked who these people were. Not getting a satisfactory answer, she independently decided to contact each person, one by one. She said that she saw they were copied on such-and-such memo and wanted to know what their role was in that project, and did they have time to talk to her about their responsibilities. Needless to say, most of the people were only too happy to meet with her and discuss their work. She now feels that by getting to know so many key people in the organization, she was given much more responsibility more quickly than would have normally been expected, just by contacting people who are copied!

YOUR WORKPLACE

We talked earlier about your workplace and its relation to "turf." Now let's talk about it as the place where you are spending a lot of your time. So, although it's not your room at home, it should be as comfortable physically and psychologically as you can make it. No, we're not going to get into the vagaries of the "cubicle" as epitomized by many memorable Dilbert cartoons by Scott Adams. But it is important to make your workspace be comfortable for you. This should mean liberal use of things that will relax you as you do your work, for example, family photos and living plants, where you put things, and how messy or orderly you choose to be. There is a common sense here, and many places to find suggestions about designing or decorating any workplace, large or (usually) small. You can get good ideas just by walking around and seeing good (or bad) examples of how others have decorated their workspaces. The point is, *it matters,* so take the time and effort to make your workspace as comfortable as you can.

"THE TYRANNY OF THE IN-BOX"

Now what is that? Well, it used to be that we had only a plain old "in-box" full of printed material—papers, memos, magazines, letters, and junk mail, and important messages from clients and your manager too!—that kept rising in height like an office snowfall. (In the spirit of full disclosure, you also have an "out-box" in which you're supposed to put all the work you finish!) Now we *also* have to contend with an *e-mail in-box* with scores of messages zooming in at all hours of the day or night from all over the world.

All these in-box messages *demand* our attention, hence the phrase, the "tyranny of the in-box." How you manage to control these incoming messages is the key to being effective. Here are some suggestions. One tried-and-true tip is to try to *handle each in-box item only once.* That is, read the message and *act* on it immediately; the only three choices are to respond to it, file it, or delete it. Try to avoid putting it aside to get to it later; this immediately adds to your "to-do" list, sapping your mental energy since you're now subconsciously thinking about when you're going to get to it. A corollary to handling something once is trying to respond immediately to the sender that you got the message. Call or e-mail them that your first reaction is thus-and-so, and you'll get back to them at a certain time. No, you can't handle everything only once, but if you are able to do this with half or three-quarters of your in-box mes-

Figure 14.1 Let technology help you manage your workplace—your "to-do" list, your calendar, and your priorities. Put your PDA to work!

sages, you'll be more effective and have increased amounts of time for your other work.

INTERRUPTIONS

What can be more fun for anyone than the prospect of talking with coworkers around the proverbial water cooler or coffee pot? There's always plenty to talk about—office gossip, current projects, last night's ball game, and many other subjects. Sure, socializing is an important part of work life, and the only caution here is that you need to *control* the amount of time you spend doing this because, duh, it can take a lot of time away from doing your work. Perhaps the most vexing aspect of this are all the interruptions from phone calls or people stopping by to talk about something. "Got a minute?" is often an innocent question, but it's *never* a minute but more like 15–30 minutes or more if you're not careful.

Phone calls are another area that can disrupt your workday. No one says that you *must* answer every phone call immediately, especially if you are in the middle of an important task or meeting—

that's why Caller ID and voicemail are such important adjuncts to our lives. Adding to the distractions are the new e-mail messages that "beep" to get your attention, and tempt you to stop what you're doing and respond (as has just happened to me as I write these very words). Well, it is OK to stop, as long as you somehow control how much time you spend away from the job at hand. You *can* politely deflect interruptions by simply telling the person you're in the middle of something now and ask when you can get back to them. And, if you should be so fortunate as to have a door, you can close it if you need the privacy to concentrate. Absent that, you can go "hide" in a conference room if you need to avoid interruptions.

APPROPRIATE DRESS FOR WORK

So, what's "appropriate" these days? What should one wear to "dress for success," to paraphrase a classic book on business attire. Just for starters, are we all clear about what "business casual" means and what to wear on "dress-down Friday"? Are business suits obligatory, or only when you're meeting with a customer or other outsider? Where is the line between acceptable and unacceptable? This is a moving target and is influenced by your type of industry, the part of the world you're in, and so forth.

The best guide is to see what others are wearing to work, find out if there are any guidelines in your organization covering dress (there probably are), and just use common sense (which, as is often noted, is not so common.) A general truism is to dress more formally or conservatively if you are not sure what to wear. And it almost goes without saying that you should adopt good personal habits of health and cleanliness, and be neat and presentable at all times.

WORK HOURS

You need to understand an important thing about work hours: your manager can control very few things about you and your work performance, but *seeing you at work* during the hours you are supposed to be there is one of the few remaining areas of control. So in this day and age of flexible work hours, shortened work weeks, and a general relaxation of the rigidities of the workplace—and, ironically, with more people working at home—being there is still important to your manager and to your coworkers. It is an unspoken rule that one has to be seen at one's desk to be "working," no matter what the quality or quantity of one's output. Even people who work from home still need to "keep in touch" with the workplace, though they are not physically there all the time.

"LUNCH, ANYONE?"

Taking the time for a relaxing lunch is often a casualty of increasingly long work hours. Yes, a relaxing lunch, not just feeding your face. "But I have no time" and "I have too much to do" are the refrains of many as they eat lunch at their desks day in and day out. This is not good, and you can show your manager this book where I said it. Of course, we need to eat, and "ordering in" is a convenience for many, but the point is that *you need to take a break from your work* and lunch is a good time for that and to meet other people—so go to the company cafeteria, go shopping, and, ideally, get out into the fresh air as often as you can. It invigorates you so you can work more effectively through the rest of the day. Eating at your desk depresses you, though you may certainly have to do it on occasion.

THE "CULTURE" OF THE ORGANIZATION

Let's conclude with a brief mention of a very important concept that also can be very difficult to crystallize. Organizations, like societies, have a "culture" because they are formed of people and conduct business in specific geographic locations. Although it is easier to contemplate the culture of a country—say, the United States or Japan—it is more difficult to identify the culture of an organization.

However, an organization's culture can be identified by "the way things are done there." This may relate to how formally or informally the people dress or act, the look of the offices, how much people interact or the amount of buzz in the company cafeteria. I recently had lunch several times in the cafeteria of a large engineering society, and was struck by the high energy and animated conversation of the people, a sign to me that these people appeared happy and that this was a stimulating place to work. As hard to define as culture might be, it is important for engineers to determine how comfortable they are in the organization—indeed, how they seem to "fit in"—as both a guide to joining an organization or staying with that organization.

The workplace is a different and ever-changing experience for each of us, and since you will spend as much of your life there as anywhere else, it is important for you to achieve for yourself a productive and comfortable work setting.

SUGGESTED READING

Bernstein, Albert J., and Rozen, Sydney Craft. *Neanderthals at Work: How People and Politics Can Drive You Crazy . . . And What You Can Do About Them.* New York: Wiley, 1992.

Campbell, David. *If You Don't Know Where You're Going, You'll Probably End Up Somewhere Else.* Allen, TX: Argus Communications, 1974.

Johnson, Spencer. *Who Moved My Cheese.* New York: G.P. Putnam's Sons, 1998.

Kanarek, Lisa. *Organizing Your Home Office for Success: Expert Strategies That Can Work for You.* New York: Plume Books, 1993.

McCormack, Mark H. *What They Don't Teach You at Harvard Business School: Notes from a Street-Smart Executive.* New York: Bantam Books, 1984.

Smith, Nila Banton. *Speed Reading Made Easy.* New York: Warner Books, 1987.

Zorn, Robert. *Speed Reading.* New York: Harper Perennial, 1991.

CHAPTER 15

DEALING WITH STRESS AND HAVING FUN

It's most helpful to hear someone telling me how to have fun. I'm a workaholic.

—Engineer at SWE seminar

"**I**'M REALLY STRESSED OUT!" Things are piling up, there's not enough time, stuff is due tomorrow, and you're tired and burned out. It's not a pleasant feeling, to be sure. But what can we do about it? Surprisingly, a lot.

First, though, let me share with you my personal turning point in dealing with stress, the low point of my career. It will lead into a discussion about what stress is from a physiological viewpoint and then what things you can do to manage stress in your life.

In the mid-part of my career, with 15 years of experience under my belt, I was put in charge of a team to review ground access to the three New York–New Jersey airports because senior management was dissatisfied with the alternatives recommended by staff. Frankly, I was in way over my head. To add to this stressful situation, my mother had passed away only a few years earlier and my father was then fighting cancer. Psychologically, I was a ticking time bomb, but I didn't know it.

I soon found out, just as I was supposed to deal with a set of major issues and conflicts among difficult, powerful people who were putting pressure on me to produce what they wanted. I began to take work home with me for the first time, to wake up in the middle of the night with worries on my mind and be unable to get back to sleep, to go to work earlier and earlier, to eat at my desk every day, and to stay late often. The worst was the weekends. For the first time, I began to stare at the clock and calculate how many more hours remained before I had to go back to work on Monday. This was all new to me. Up until this time, I had enjoyed all of my work, even some nasty and unpleasant assignments.

This situation came to a head about four months into this difficult project. In rapid succession, the manager I reported to told me to my face: "Carl, It doesn't look like you're doing much here." It felt like a trap door had opened beneath me. Next came a meeting of the senior task force that said pretty much the same thing. And right after that meeting I called my father's doctor to get the message, "Mr. Selinger, your father has expired." I sat in my office stunned, and then I needed to escape. I got through all of this, as we all must do at times—the work pressures, the shock of dealing with parents passing away, and such. The lesson for me—and shared with you here—is how I learned how to better handle stress soon after this turmoil.

Dealing with Stress and Having Fun

- The body *can* deal with stress: "Fight or flee!"
- Stress becomes a problem when it adds to your load.
- Reduce stress by treating yourself, exercising, and "telescoping."
- "Potential problem analysis":
  - *What* can go wrong?
  - *How likely* is it to go wrong?
  - *How serious* would it be if it did go wrong?
  - *What would you do* if it went wrong?
- *Have fun!* Are engineers the only people that need to be *told* to have fun?
- You know what fun is for you, so make time for it. "*Just do it.*"

I went to a bookstore during a lunchtime and browsed through books on dealing with stress. There were dozens of titles. I started looking through several, and applied my technique for determining if a book is worth buying: first, I look at the jacket and reviews; second, I peruse the table of contents; and third and most important, I read a short excerpt on some issue I feel I already know about to see how the author covered that issue. If the author does that well, then I'm interested in buying the book. After going through a few books like this, I found the one that I bought and that turned around my life: *The Work–Stress Connection* (see Suggested Reading at the end of the chapter).

The first thing it said (and I'm sure many books on dealing with stress say the same things) was that *stress is normal,* that our bodies are *designed* to deal with a certain amount of stress. The most basic way to think about this is to consider the "fight or flee" response we exhibit when we confront a stressful situation. We're not too far from the days when we would be out in the woods and encounter a lion or a bear (this can still happen). We have two choices to make instantaneously: stand there and fight, or flee as fast as we can.

> I need to keep stress levels lower by understanding possible outcomes.
> —Engineering student, Webb Institute

The problem in the "real world" is that we often can't do either, and that's where stress starts to become harmful. You just can't fight your manager, co-workers, or customers; it's not socially acceptable to do this. Of course, you can argue, raise your voice, assert yourself, disagree, and so forth, but you cannot really say what's on your mind without facing serious consequences of being fired or worse. And you also can't really run away from the situation and hide, as much as you may want to. Now, you're wrong if you think I'm going to give you a list of things to do here. You'll need to do this by applying your people skills to the situation. But recognizing how your body is reacting to a confrontation with a stressful situation is an important first step.

> I don't know if I'm prepared for the "real world." It's easy to be afraid of the unknown.
> —Engineering student at a Tau Beta Pi seminar

The next thing to realize is that our bodies can deal with a certain amount of stress. The book I read said it was like dragging a load of stones. You're strong enough to drag up to a certain weight of stones, and then it becomes increasingly difficult as more stones and weight are added, up to a maximum point at which you can't take it anymore. This is the way our bodies deal with stress. We can handle it up to a certain point, and each of us has different abilities to handle amounts of stress. So what can we do to manage stress in

Figure 15.1 We often are under stress to finish things by a certain time and have too many other things on our minds.

our lives and reduce the harmful aspects of too much stress? At the same time, it is arguably healthy to have a certain amount of stress in our lives to keep ourselves sharp, "on our toes," and have an edge.

The book I read, and subsequent things I've learned, provide at least three things we can do to manage stress in our busy lives: exercising, treating ourselves, and a technique called "telescoping" (which is different than the telescoping method discussed in Chapter 6 to quickly sort priorities).

EXERCISE

Getting exercise is one important way to manage stress in your life. No, I am not going to launch into why you should start a diet to lose weight, or get a personal trainer, or buy the latest exercise equipment you saw on an infommercial (that you buy for your home and neglect to use after the initial enthusiasm wears off). Rather, you need to understand that *exercise offsets the stress in your body*; it's like the "flee" option. Whatever exercise you do will directly lessen your stress. When you were young, you probably were very active, or at least more active than you may be now. So your body was able to better fight the stress of exams and papers and other pressure situations you faced in engineering school. But since school, you've gotten into a complex real world of more work to do, perhaps a family you need to support, and situations you must face—all adding to the weight of stones you're trying to drag around. So the more you exercise, you will benefit by a direct reduction of the stress you feel, and be better able to deal with the situation.

TREATING OURSELVES

Treating ourselves also works to reduce stress. Frankly, I can't tell you *why* this works, but it does for me. When I first read that book I bought, it said to buy myself a present during lunch, say a tie. Since a lot of stress is feeling badly about yourself for not being able to handle things, why not buy yourself a present because you're really a pretty good person. Give yourself a reward. Cut yourself some slack. So I bought a new tie at lunch after reading this, not expecting much, and I actually began to feel better!

A related thought fits here. You may be stressed because you feel you *should* be doing this or *should* be doing that. Instead, try thinking, "Don't *should* on yourself!" Get it?

TELESCOPING

Telescoping is a wonderful technique that I have used repeatedly, almost every week in fact, to help my body reduce the stress of upcoming activities. For example, you're going to be running an important activity soon and you're afraid it's going to be a disaster, since one of "Murphy's Laws" is that "Whatever can go wrong, will go wrong." This is a very normal, pessimistic "worst case" attitude that we all tend to assume, but this way of thinking is *not helpful.*

Telescoping is an easy form of "contingency planning" that you can do in your head most of the time. First think of the upcoming event or assignment that you're worrying about. OK, do you have something in mind? Now assume the worst. That should be easy because that's probably what you usually do, and that's what is causing the worry. So be very pessimistic. Think of how bad it could be, and—this is important—what could be the consequences of the worst happening. Could you get fired, severely reprimanded, lose a customer, told to "try better next time"?

. OK, now think of the entirely opposite outcome, that the thing goes very well. Be optimistic, visualize what could happen: being praised for doing a "great job!" by your manager, co-workers, or customers; or getting a commendation or promotion. Now for the third and last step—think of what a middle outcome might be—it goes pretty well and you don't mess up—which would probably be the most likely outcome. So what's the value of "telescoping"? You and your subconscious have now evaluated the extremes of the upcoming situation and, most importantly, have found that the worst case, though not being something you wish for, would most likely not result in something so terrible. And, in fact, you may now be in a better frame of mind to achieve a more positive outcome simply by doing this mental exercise.

POTENTIAL PROBLEM ANALYSIS

There's a similar technique that can be useful and applies a bit more rigor, called "potential problem analysis," drawn from the Kepner–Tregoe program (which also includes valuable modules on "decision analysis" and "problem analysis"). Let's take the same example we were thinking about before. Now think, in sequence: *What can go wrong? How likely* is it to go wrong? *How serious* would it be if it did go wrong? *What would you do* if it went wrong? This step-by-step approach can almost be done in your head, and is very effective.

Let me give an example of where "potential problem analysis" worked for me. I was running a boat ride for about 300 attendees of an engineering conference in New York City, and we chartered the famous Circle Line, the extraordinary boat ride around the island of Manhattan. We had made all the arrangements one had to make, and, except for guaranteeing good weather, we were good to go. During the week before the boat ride, I did a "potential problem analysis" on a blank sheet of paper. I wrote down all the things that could possibly go wrong, and had a list of about a dozen items, one of which happened to be that the "Band doesn't show up." We had hired a band for entertainment, recommended by one of our members. The next step in the analysis was "how likely" was it that they wouldn't show up? It was unlikely; we called them the week before and all was fine. The third step was "how important" was it that they show up; and it was moderately important, since the band was our entertainment and we wanted to let people dance. The fourth and last step was "what would you do" if it happened.

At this point, you might be saying to yourself, "Well, this will never happen, so why bother thinking of what you would do for every possible thing?" But I persisted and thought about it, and decided that if for some reason the band had not shown up by ten minutes before departure (they were due to arrive about an hour beforehand), then I would hop in a taxi and go to a specific music store in nearby Times Square, and ask the taxi to wait while I bought a radio (a "boom box" to be more precise) and a bunch of tapes (no CDs then!), and that would be our music. You'll never guess what happened! The band did not show up (and we never found out why, but we saved their fee!) and just as I was ready to go get the music, one of the hands on the Circle Line said he had a stereo and tapes and could patch the music into the boat's public address system. That's what we ended up doing, but the "potential problem analysis" anticipated that situation, and many others, and allowed me to deal more comfortably with managing the stress of running a complicated event.

HAVING FUN

It seems very ironic and, at the same time, totally appropriate, that one of the last "skills" we cover in this book is the need for engineers to learn how to have fun. Yes, having fun is a wonderful part of our lives and also an effective way to deal with stress. And yet many of us have trouble finding time to have fun, or, when we do, feel guilty that we are enjoying ourselves when there are more important things to do.

My son Doug, a scientist, put it well when he saw that I urge engineers to "have fun." "Are engineers the only people who need to be told to have fun?" Well, perhaps yes. To be sure, many readers are saying that this doesn't apply to them, that they have plenty of good times mixed in their lives, and that is fine. But many engineers

Figure 15.2 Reading for pleasure is very relaxing and you can also learn a lot. And, yes, you can make time to read!

get into a workaholic mode that is drilled into them in the "boot camp" of engineering school, where they can be overwhelmed with the amount and difficulty of the work, and are forced to work many more hours than students in other fields. This carries over to the workplace when the hours increase and there is less time for personal business, relaxation, and having fun. Recent statistics show startling increases in hours worked per week by people in the United States, resulting in less personal time.

Your son saying "engineers are the only people who need to be *told* to have fun" is *exactly* true. Completely dead on. That was probably my favorite part of your seminar. It opened my eyes to explaining the way engineers behave. My dad, who is an engineer, is a complete workaholic. Having fun seems like it is a chore for him. Your son's description is so accurate it's not even funny. Make sure you always include that in your presentation.

—Senior chemical engineering student at Cooper Union

Before I "teach" you how to have fun, let me tell you a story of a junior in engineering school who had been in a communications workshop I gave and e-mailed me afterwards to confide about some problems she was having. Linda's first message was pretty serious—she was having a number of problems with the heavy workload and her roommate, and her mother was very sick. All this was really stressing her, especially the unrelenting workload. I tried to encourage her to stick with things, talk to her professors to get some relief—the usual advice. I also told her to take some time for herself to go have fun. In a series of messages over a week or two, Linda seemed to be having more and more trouble, and I was ready to refer her to get some professional help.

At the same time, I kept urging Linda to take more breaks and have some fun. While it would not be the whole answer to her difficulties, it would be a way for her body and spirit to be refreshed and relaxed a bit. I asked her what she did for fun, and she said that she loved the movies, but did not have any time, and would feel very guilty if she went to a movie while she had all her work to do. This came to a head one day when her e-mail messages sounded increasingly troubled. I asked if it was OK for me to call her on the phone to talk about things. Our conversation was serious and I tried to suggest ways to handle her workload, her roommate situation, and then how to have some fun to relax. On that she was adamant—she had no time for fun. So, I said that I was a professor at her school too, and even though she was not in my class, I *assigned* her to go to a movie that very day. As it turned out, she listened to me. I can't say that was a turning point in Linda's life, but she did get better and graduate and is now doing well as an engineer at a consulting engineering firm, but it shows how hard it is for some of us to enjoy life. Which begs the question: *What is fun for you?* Actually, this is an easy question, since you know already know what fun is for you; you just need to make time for it, or, to use the famous Nike slogan, "Just do it." But let's talk about this for a moment, since having fun can be serious business.

The easy part is that you already know what you enjoy doing. The gamut runs from playing sports to reading, going to the movies, hanging out with friends, listening to music, and so on. I always ask attendees of my seminar to share out loud what's fun for them, and the list is always long and varied. One time, a young engineer said that he read electrical engineering textbooks for fun. He got a huge laugh from the others, but he was serious, so go figure.

Often, the hard part for many of us is *finding the time* to do what we enjoy. Without getting into deep philosophical treatise on managing your time (actually, we did this in Chapter 6 on "Setting Priorities"), you simply need to *make* time for having fun. This means looking at your calendar and fitting things into your busy schedule. Even go to the point of scheduling "appointments" to do fun things, if it comes to that. This is so basic that it hardly needs examples, but let's give some examples anyway. When will you get to the gym to work out? Find specific slots in the mornings before work, at lunchtimes, on certain days after work, or on weekends— and stick to them. Are you working until 8 pm every night? Find someone to hang out with for a late dinner once in a while. Do you have meetings from morning to noon to night? Then plan 15 minutes to grab coffee with someone at one of the meetings, either before, in between, or afterward.

Let's not forget taking a *vacation,* a time meant for fun and relaxation. When we're young and right out of school, taking a vacation seems not so big a deal. We're tired of school and want to get to work, until we find that someone is paying us to work all year, and getting only two weeks or so off for a whole year is suddenly very constraining.

Both of the following stories are similar. I was giving a seminar at the University of Connecticut and asked the group what important decisions they had to make (you'll remember this from back in Chapter 4). When I went around the room and asked people to share what decisions they identified, a professor first demurred, saying his decision wasn't very important. I persisted and asked him to share it with the group if he felt comfortable. He then said that he was trying to decide if he should take a vacation that year. I asked him why this was an important decision, and he said his wife was urging him to take a vacation because he hadn't done so for about five years. I was taken aback . . . and immediately replied to him, loudly, in front of everybody: "Yes, *take a vacation*! That's your decision!" Then I launched into the importance of getting relaxation for yourself, spending time with family, and so on. I am happy to say that he got the message.

The other vacation story involved a co-worker, John, who was rather proud of the fact that he hadn't taken a vacation for many years, that he could never find the time to get away. Year after year, nothing I could say to him changed his attitude on this, until one year the memo came around in the spring for everyone to select the weeks they wanted to take vacation that summer so schedules could be coordinated. I remember confronting John at the copying machine and asking when he planned to take vacation that summer. He stammered that he didn't know, and that he never took vacation because he had many projects he was working on. So I said, point blank, "John, Go back to your office *now,* look at your calendar *now,* pick two weeks for vacation *now,* and I will help you manage any work tasks when you're out." John said "Really?" I said "Yes," and he did what I suggested—took vacation that year, had a great time, and told this story to many people for years afterwards about how I got him to take a vacation. Go figure!

Managing stress and having fun are such important parts of your life. They can be done, but you need to make the effort.

SUGGESTED READING

Davidson, Jeffrey P. *The Complete Idiot's Guide to Managing Stress.* New York: Alpha Books, 1997.

Skye, Paul. *Mastery of Stress: Techniques for Relaxation in the Workplace.* St. Paul, MN: Llewellyn Publications, 1998.

Veniga, Robert, and Spradley, James. *The Work–Stress Connection.* New York: Ballantine, 1981.

CHAPTER 16

TAKING ACTION AND SUMMING UP

Fear stops us from doing many things, often for many years. Yet we have to act in the face of fear and when we do, we often find out it's easier than we thought. So fear is in our heads. Taking action gets us out of our heads.

—Engineering manager, TransCore

To PARAPHRASE A favorite saying of authors at this point: If you're still reading this, then please understand that *anything you have learned about nontechnical skills will only be valuable if you take action in your life.*

You've probably heard the advice about the importance of being "action oriented" or "proactive." There is fundamental truth here: you must adopt a personal style of tackling things rather than being dominated by the usual excuses to put things off, procrastinating, fearing the consequences of your action, waiting until everything is "perfect," or waiting for someone to tell you what to do. A ship under power is going places; a ship floating adrift is subject to many bad things happening to it.

Let me share a story that happened only a few years ago, well into my career, but which suggests that you should always be learning things, and in this case provides a helpful way for you to be more action oriented. I was speaking to my manager, Gary, about an idea for an airport initiative. He liked the idea and then asked what my "action plan" was to do it. Action plan? I hadn't thought about that, but told him I'd get back with one shortly. Some time later, the scene repeated itself: I suggested another idea to Gary, who again was supportive and again asked for my action plan. This time, I had

Figure 16.1   Take action! Get on the phone or e-mail someone to follow up on a project, deal with a problem, or say "Thank You"! Don't procrastinate!

a specific set of actions in mind because I had prepared for the question.

Broadening this, it will be a very valuable skill for you to think in terms of an "action plan" for many things in your life, certainly projects at work, as well as issues you need to tackle in your personal life.

I challenge you to take a moment to write down your personal action plan on this simple template for things you need to tackle right now at work or school and in your personal life.

Your Personal Action Plan

- What I will do when I go to work/school _____
- What I will do when I get home _____

I'm sure you get the idea. Your action plans could be for specific projects, for the coming week, for personal development, anything that helps you focus on the things you need to do. You can put your action plans on paper (or on a Word document on your computer or PDA) or just keep it in your head, whatever works best for you. *The important thing is that you commit yourself to taking action.* So act on the important things in your lives!

> **Take Action!**
> - Make a personal action plan:
>   - What I will do when I go to work/school
>   - What I will do when I get home
> - Now act!
> - It's what you do, not what you say!

I recently met with several educators at a major engineering school and discussed my seminar and the various nontechnical skills that are covered. When we got to this subject of taking action, and developing one's personal action plan, one of them asked me, "But Carl, how do you ensure that people actually *do* their action plan?" I was a bit startled by the question, though in retrospect it reflected the mindset of the academic community that is focused on students turning in required homework and papers for grades. I replied that I *can't* ensure that they do their action plan. It is up to them to follow through. It is their personal responsibility to act, for their organizations and for themselves. So, I say to you that it is up to you to take responsibility for your life, and a key part of this is your acting on the things you decide to tackle.

> There are many ways of going forward but only one way of standing still.
>
> —Franklin D. Roosevelt

Let me repeat a story that may help you to deal with actions involving unpleasant situations. A few days after doing a seminar at Bucknell University, I received a very satisfying e-mail message that went something like this: "Carl, I listened to you telling us that we have to act and to tackle the most important things on our plate. I work part-time while attending school and had been putting off contacting a customer about a serious problem. This was really bothering me to the point where I was losing sleep over it. So I resolved to listen to your advice and the first thing the next morning

> Significant (effective) action can occur in very little time. It could be a three-minute conversation, a quick e-mail, a phone call, and so on. Epiphanies (transformations) can occur in literally instants and overcome years of being stopped (often by fear).
> —Engineering manager, TransCore

I sucked it up and called the person. It turned out the situation was not that bad and he understood. I was so relieved! Thank you!"

To be sure, taking action does not mean that all your problems will be resolved with one phone call or e-mail message. It does mean, however, that you will be more in charge of your life. If you act with determination, you'll solve the small problems before they become big and "bite" you.

> I need to walk the walk . . . stop talking and just do it.
> —Engineer at a major Midwestern engineering company

Let me give an example of how the idea of being action-oriented is reflected in pop culture, in this case, a recent popular song. So, in the words of one of my favorite singers, John Mellencamp, "It's what you do, not what you say. If you're not part of the future, then get out of the way." Your actions will speak louder than words, but you need take these actions. You need to make your own future, and you will feel better about yourself for having taken the initiative. End of the pep talk!

So what's left to say? You're probably graduating from engineering school, or you've found a good job and been out in the engineering profession for a number of years. Congratulations! But have you felt well prepared for life as a working engineer? Do you know, for example, what steps to take to advance in your career? Do you know how to stay current and competitive in your field? Can you deal with difficult people, like your manager, or clients, or the public? Are you comfortable speaking in front of a crowd?

> I feel it is very important to be exposed to this "stuff" early in one's career. Thank you!
> —Engineering trainee, Port Authority of New York and New Jersey

These are the areas we've tried to cover in this book. If these issues concern you, please gain some measure of comfort from knowing that *you are not alone.* You have only to read the concerns about the "real world" from young engineers voiced in the "boxes" throughout the book (and also in the Appendix). The fact is that many, if not most, young engineers emerge from school with fabulous technical talent, but little ability in these "soft" skills. I can't blame the engineering schools for not covering this ground effectively; after all, they have their hands full just educating students in the latest technology. But when those students eventually enter the workplace, they may find their soft skills woefully undeveloped. Who will teach them then? That's been the goal of this book: to acquaint you with the most important nontechnical soft skills that each engineer needs to be more effective in the workplace and happier in life.

I have learned that the outside world actually craves a well-rounded engineer. An engineer who knows how to manage his or her time and participates actively at the place of work and, while doing all this, can have some fun.
—Chemical engineer at AIChE seminar

After all is said and done, I believe that the "bottom line" is that it is up to each of us to *be the best person and professional that we can be*! We are all endowed with certain basic talents and abilities, and it is then up to us to develop our skills and apply them effectively in the "real world" in which we live and work.

It forced me to think about things I don't usually think about—what's important to me, and so on.
—Engineer at ASME seminar

If it has not been apparent by now, I confess that I am a card-carrying optimist. I believe that the glass is always half full, not half empty. (And there is certainly some truth in the joke that engineers would say the glass is overdesigned!) Despite the many terrible problems in the world today, we live in a wondrous time in which technology is transforming the world we live in, and we are technology professionals who will influence these applications for the betterment of society. Much of what you will accomplish in your life is under your control. So continuously build your education and skill set, always be willing to work together with other people, keep up your technical skills, and apply the soft nontechnical skills you've learned in this book.

> I'm concerned that I will be comfortable with my skills/weaknesses, know my limits, and know how to work my best every day.
> —Engineering student, Binghamton University

If you do all this, I believe that you will be a more effective and happier person. Now go and do, and good luck to you all! And it would be wonderful to hear from you about your successes!

> But all of this is easier said than done.
> —Engineer at Tau Beta Pi seminar

SUGGESTED READING

Bliss, Edwin. *Getting Things Done*. New York: Bantam, 1980.
Foster, Richard. *Innovation: The Attacker's Advantage*. New York: Summit Books, 1986.

APPENDIX

1. MANAGER'S SURVEY ON NONTECHNICAL SKILLS OF YOUNG ENGINEERS

Surveys have been done with engineering managers from several organizations to find out *how important* it is for young engineers to possess each of a dozen specific nontechnical skills, and to obtain the managers' ratings of the *perceived abilities of young engineers* in each skill area. Below are the tabulated results of the survey conducted at one major engineering organization.

Organization X
Engineering Managers' Survey of Nontechnical Skills of Young Engineers

Ten managers rated the importance of specific nontechnical "soft" skills to young engineers, and their perception of young engineers' abilities in each area, related to their career success and the organization's effectiveness. Managers were also asked what critical issues (problems and opportunities) they saw in developing the nontechnical skills of young engineers, and if they had any other suggestions.

Findings

General-to-strong support for this type of training in nontechnical skills. The larger the arithmetical "gap" between ratings for "Importance" and "Level of ability," the higher the "need." Highest need: writing (4.0), teamwork (3.1), public speaking (3.0), and decisiveness (2.5). High need: interpersonal skills, meeting effectiveness, and taking responsibility (all 2.3). Lowest need: negotiating (1.2), ethics (0.9), and risk taking (0.6).

Stuff You Don't Learn in Engineering School. By Carl Selinger.   **157**

Survey

Please rate from 1 to 10 (10 being the highest) your impression of the importance of the following areas to the effectiveness of your organization and to the professional success of engineers. Also rate the level of ability you see in young engineers in your organization.

Skill	Importance? (10 = very important) (1 = nice to have)				Level of ability? (10 = excellent) (1 = inadequate)		
Tabulation →	(Lo-Hi)	Σ	Avg.	"Gap"	(Lo-Hi)	Σ	Avg.
In order of priority:							
Writing	(8–10)	89	8.9	4.0	(3–7)	49	4.9
Teamwork	(9–10)	97	9.7	3.1	(5–9)	66	6.6
Public speaking	(3–10)	75	7.5	3.0	(4–7)	45	4.5
Decisiveness	(5–10)	81	8.1	2.5	(3–8)	56	5.6
Effective at meetings	(6–10)	78	7.8	2.3	(3–8)	55	5.5
Taking responsibility	(7–10)	86	8.6	2.3	(4–8)	63	6.3
Interpersonal skills	(7–10)	85	8.5	2.3	(3–10)	62	6.2
Creativity	(5–10)	77	7.7	2.2	(4–8)	55	5.5
Leadership skills	(4–9)	72	7.2	2.1	(4–8)	53	5.3
Setting priorities	(5–10)	79	7.9	1.9	(4–8)	60	6.0
Negotiating	(1–7)	45	4.5	1.1	(1–7)	34	3.4
Ethics	(5–10)	88	8.8	0.9	(5–10)	79	7.9
Risk taking	(1–8)	49	4.9	0.6	(2–8)	43	4.3

What critical issues (problems and opportunities) do you see in developing the nontechnical skills of young engineers?

- It is a great plan [to have training for nontechnical skills]. Concerning budgets, the more the incoming young engineer has, the better. Also, this program can help retain young engineers because the engineer may look at the training as the company wanting them to move up within the company.
- Negotiating, writing, public speaking, and creativity.
- Impatience on their professional development. Want to be promoted to a high position quicker and make more money than their capacity.
- Communication skills (writing, public speaking, etc.) seem in general sorely lacking with young engineers. It seems more and more colleges are focusing on technical education and minimizing the liberal arts portion of the curriculum. While the technical end of this profession gets more and more "complicated" with

advances in technology, effective communication becomes more demanding. Higher percentages of times are now spent on communication and report writing and often I have seen poorly written reports lacking coherence.

- Writing and public speaking are the two most important soft skills that engineers need to succeed in the field today. There are many engineers who can perform the technical work, but being able to convey that technical work to the client and/or public is what allows engineers and the company they work for to prosper.
- They have to like what they are doing. They have to understand that communication with others is very important.
- Good seminars/training shall be provided in technical writing, interpersonal and decision-making skills. From time to time, management shall discuss the benefits of teamwork with the young engineers.
- Working effectively and efficiently as a team is very important for a young engineer to learn. Working alone can result in significant re-work if scope is not clearly defined.
- To provide *consistent* opportunities where skills can be utilized.

Are there other issues you think are important, or other suggestions about this topic?

- This sounds like a great internal workshop that should be advanced. On-the-job training does work but if each young engineer had this type of training (such as PM Boot Camp), we will be that much more ahead of the game in developing future Project Managers.
- More training seminars provided to the engineers should help them to improve the abovementioned issues.
- As a Technical Dept. Manager, I am more concerned about developing their technical skills but it is nice to see them have a good understanding of the engineering business.
- How to convey and react to mistakes by others team members? How to react to mistakes conveyed by other team members and seniors?

2. ENGINEERS' CONCERNS WITH THE REAL WORLD AND OTHER ISSUES

Engineers attending my seminars over the past decade have indicated their feelings about the various soft skills, what they need to develop the most, and what most concerns them about the real world. Below are selected responses, many of which appear in "boxes" throughout the text. The responses are organized by chapter headings, though many of them cover many skill areas. They express unique and varied insights and feelings, and can be used to stimulate discussions on professional development programs for practicing engineers and engineering students.

Chapter 1   Stuff You Don't Learn in Engineering School

- "They definitely don't teach this [stuff] in school. It's not something I would have heard elsewhere." Engineer at Tau Beta Pi seminar.
- "It made me realize that it is not good enough to be technically competent. One needs other skills to succeed." Engineer at ASCE seminar.
- "Generally, the real world requires a person to be a well-rounded individual. Lacking a certain skill puts us at a disadvantage." Engineer, major construction company.
- "It is very important to get an idea of how things really work outside of school, and hearing about this stuff would have been helpful to me when I graduated." Business manager attending Tau Beta Pi seminar.
- "What exactly is the real world? If you're talking about whether college has prepared me professionally, technically, and academically for the 'real world,' then I would have to say, not entirely. I don't think I could ever be prepared for it unless I actually experience it first-hand." Senior engineering student.
- "I need to understand how to be responsible for myself and my life." Engineer at a major Midwestern engineering company.
- "What you don't know *will* hurt you and hold you back." Consulting engineer at a Cooper Union seminar.
- "Will I grow along with the changes in technology?" Engineering Trainee, Port Authority of New York and New Jersey.
- "Reminder of nontechnical skills of engineering. Reality check to myself that I need to work on certain things." Engineering Trainee, Port Authority of New York and New Jersey.
- "The development of these skills must be balanced with technical development. Recent trainees that we have hired have great computer skills, which are critical to developing designs and drawings. The trainee program should include seminars for lead-

ership training, public speaking, presentation, and technical writing." Assistant Chief Architect, Engineering Dept., Port Authority of New York and New Jersey.

- "Engineering schools do an inadequate job at producing well-rounded engineers with background courses in English, writing, economics, history, and so on. The trend also appears to be against this, that is, more technical and management. The latest push industry-wide toward "continuing professional development" also emphasizes the technical side, that is, they are not counting CEUs in nontechnical courses. Therefore, the burden is going to fall on the employers, who, by and large, are engineers with the same weaknesses." Engineering Manager, Port Authority of New York and New Jersey.

Chapter 2   Writing

- "Communication skills (writing, public speaking, etc.) seem in general to be sorely lacking with young engineers. It seems more and more colleges are focusing on technical education and minimizing the liberal arts portion of curriculum. While the technical end of this profession gets more and more 'complicated' with advances in technology, effective communication becomes more demanding. Higher percentages of times are now spent on communication and report writing and often I have seen poorly written reports lacking coherence." Engineering manager, DMJM+ Harris.

- "Communication comes in many forms in life—both in the workplace and at home. Meetings, memos, speeches, even simple conversations at lunch, all require strong communications skills to convey the proper message. Yet, engineering schools seldom teach such fundamentals." Mechanical engineer at ASME seminar.

- "Something good has come out of the computer revolution. The 'spell check' eliminates the worst of grammatical, punctuation, and spelling errors. Now, if they only knew how to write." Engineering manager, Port Authority of New York and New Jersey.

- "I need to develop my writing and speaking . . . to gain the ability to speak what I believe freely, not worrying about the negative impression it could possibly give people." Engineering student, Webb Institute.

Chapter 3 Speaking

- "Writing and public speaking are the two most important soft skills that engineers need to succeed in the field today. There are many engineers who can perform the technical work, but being able to convey that technical work to the client and/or public is what allows engineers and the company they work for to prosper." Engineering manager, DMJM+Harris.
- "Try to convince young engineers that it's OK to ask questions." Chemical engineer at AIChE seminar.
- "I need to develop my speaking skills, speaking up in meeting situations. Generally, I am not assertive enough." Engineer at Rice University seminar.
- "I think I need to most develop my writing. I always worry that my e-mails don't efficiently convey my thoughts." Engineer at Tau Beta Pi seminar.
- "I need to develop my speaking skills in formal presentations. I'm comfortable speaking in front of a large group of people in an informal setting, but when I have to formally present something, I get pretty nervous. No matter how much I prepare, I don't feel like I prepared enough." Engineer at Tau Beta Pi seminar.

Chapter 4 Making Decisions

- "I'm *afraid* to make decisions!" Engineering student, University of Connecticut.
- "You don't realize how many decisions you need to make every day until you really think about it." Engineer at DMJM+Harris.
- "I find myself tending toward "Well, I'm not sure, what do you think?" instead of just making a decision right away." Senior engineering student.
- "I need to make decisions, whether something good or bad that comes out of it. Also, health is a priority." Engineering senior, Binghamton University.
- "I'm glad to learn about how to make decisions. I'm glad I'm not the only one who has problems making a choice." Engineering student at Tau Beta Pi seminar.
- "I recognized several flaws in my decision-making processes that needed to be corrected. Particularly, I realized that putting off a decision involves making a decision in and of itself." Civil

engineering Trainee at Port Authority of New York and New Jersey.

- "I need to develop my decision-making skills. Engineers are trained to *optimize* decisions. Gathering data can become the trap to *making a decision*. Not because of fear, but because we need all the data (or so we think). We need help to break that cycle." Engineer at major Midwestern engineering company.

- "I feel better now if I have to make any decision because there is no right or wrong decision. All I need is to get enough information to help me in making the decision." Chemical engineer at AIChE seminar.

- "The most helpful part of making decisions is not being *afraid*!! I fear the consequences but I can't look back." College engineering senior at SWE seminar.

- "It was helpful being forced to think of the decisions that I have to make." Senior Cornell engineering student at SWE seminar.

Chapter 5   Getting Feedback

- What concerns me most about the "real world" is the lack of feedback from most all of the people I work with. In school, I received grades. Here, it's nothing, unless I make a mistake, it seems." Engineer at major Midwestern engineering company.

- "Feedback is motivating, but too infrequent." Engineer at Cooper Union seminar.

- "I ask for feedback but I often reject it as 'He doesn't know anything' and then don't utilize it for my betterment." Engineer at DMJM+Harris

- "What we don't know can hurt us. So it is important to request feedback from others. What we can't see in ourselves, they may be able to see. Once it's revealed to us, we can do something about it." Engineering manager, TransCore.

Chapter 6   Setting Priorities

- "I have seen 'setting priorities' in practice being a problem for most people coming out of school." Engineer at Rice University seminar.

- "I've been so busy that I haven't really had time to sit down and think about where I'm going in my life and what my priorities are right now. I need to stop and think more about my priorities." Engineering student at Tau Beta Pi seminar.

- "I have started to prioritize my life a little better. Unfortunately, you were right about our health being the top priority. My body decided it would no longer take all this abuse I've been putting it through. I became ill and now I'm just catching up with all my work." Engineer at Tau Beta Pi seminar.

- "I'm worried about whether I'm going to know how do to everything I'm asked to do." Engineering student, Binghamton University.

- "I need to think about what decisions are really important, and to distinguish between the small details that tend to take over and start to dominate my decisions, and what is really important." Engineer at Tau Beta Pi seminar.

- "I need to develop my ability to prioritize. Since I like to procrastinate, everything that is 'important' becomes 'urgent'." Mechanical engineer at ASME seminar

- "Now I am actually making a to-do list every day. I've always had an organizer, but I've just never used it so much. I've just tried to rely on my mind more. Obviously, that wasn't working for me." Senior engineering student at Tau Beta Pi seminar

- "I'm concerned about prioritizing my personal with my professional life; how to put them together and manage them together." Engineering Trainee, Port Authority of New York and New Jersey.

- "I'm concerned about keeping up with technology, competing with other people for everything. People are always overworked. I only want to work 40 hours a week but it's hard when so many others are willing to work 60." Engineering student at Rice University.

- "I wish I had learned this when I was a student! When I was a young professional, I struggled with the differences between urgent and important, and responsibility and accountability." Engineering manager at AIChE seminar.

- "Professionals learn these soft skills as they go. However, if they learned this stuff as students the 'growing pains' of professional life would be a little smoother." Engineer at AIChE seminar.

- "I've been so busy that I haven't really had time to sit down and think about where I'm going in my life and what my priorities are right now. It made me stop and think more about my priorities." Officer of Tau Beta Pi student chapter.

- "I'm most concerned to ensure that my priorities are aligned with the priorities of my superiors." Mechanical engineer at ASME seminar.
- "Sometimes it's difficult to perceive when given particular instructions by a supervisor. I've learned to 'use ESP' with some supervisors." Engineer at DMJM+Harris.

Chapter 7   Meetings
- "Perhaps the only situation involving oral communication that people dread more than the presentation is the meeting. For many, the Dilbertian image of one person droning on endlessly while everyone else is asleep (or wish they were) is all too real. Yet, meetings do not have to be that way." Mechanical engineer attending ASME seminar.
- "What concerns me most about the 'real world' is how much time people feel they waste in meetings." Mechanical engineer at ASME seminar.
- "We must use an agenda during meetings. Most meetings I attend are for topics with no agendas." Mechanical engineer at ASME seminar.
- "Certain practices are essential for good meetings, especially asking why should I go and what do I need to contribute." Engineer at major Midwestern engineering company.
- "If everyone followed this advice, meetings would be much more useful and less painful." Engineer at a major Midwestern engineering company.

Chapter 8   Understanding Yourself and Others
- "I'm concerned that I am too self-involved in my work. I take things too personally. Also it's very difficult to deal with management that doesn't seem to listen." Mechanical engineer at ASME seminar.
- "What concerns me the most is how to deal with professional people and have to prove to myself that I'm really on the right path doing the right thing in the right place." Young engineer at AIChE seminar.
- "How do I not be a perfectionist?!" Engineering student at Tau Beta Pi seminar.
- "I have an inability to get started and/or complete projects when I don't see any relevance for them. I also have difficulty starting new things." Engineering grad student at MIT.

- "I have to learn to speak up and ask questions when I need help. I can't worry about what someone else will think." Engineer at Tau Beta Pi seminar.

- "No matter where you work or wherever you are, people are the same and we deal with similar situations. It is a matter of how we deal with them and what type of attitude we start with." Mechanical engineer at ASME seminar.

- "I learned that most engineers experience the same problems no matter what the company. That's the way business is run today." Mechanical engineer at ASME seminar.

- "There is no sure-fire method for successful communication. Each of us has to discover what works for us. The only way to discover what works is to interact with people." Mechanical engineer at ASME Hartford seminar.

Chapter 9   Working in Teams
- "I need to be able to explain to other people about the basic principles of teamwork. Some people around here do not know how to involve you in the big picture." Engineer at a major Midwestern engineering company.

- "Working effectively and efficiently as a team is very important for a young engineer to learn. Working alone can result in significant rework if scope is not clearly defined." Engineering manager, DMJM+Harris.

- "I need to work more on teamwork and working with others, as I am a very individualized person." Senior engineering student, Webb Institute.

- "Being a better communicator and more of a people person when it comes to interacting with the engineering team and client cannot be overemphasized in terms of success and job satisfaction." Engineering manager, Port Authority of New York and New Jersey.

- "Do people (team members) know they are responsible for their work even though you are the head (CEO) of the team?" Engineering student, Binghamton University.

Chapter 11   Creativity
- "How do I come up with new things or look at things in new ways without 'reinventing the wheel' or otherwise wasting time?" Engineering student at Tau Beta Pi seminar.

- "I need to be more creative. I tend to get stuck in the proverbial 'rut'." Engineer at DMJM+Harris.

- "The most important character traits that I would like to see more of are ambition, enthusiasm, and proactive thinking. Engineers coming out of college seem to look around to the experienced staff around them and seem to want to follow by example more than lead. I would like to see more creativity in their thinking and more expression of their own ideas in problem solving. I would like to see more enthusiasm in tackling and finishing each project assignment and being anxious to apply what they have learned to their next project. This relates to their own confidence in their interpersonal skills and in their own abilities." Engineering Manager, Port Authority of New York and New Jersey.

Chapter 12   Ethics
- "Responsibility, integrity, and the like are important. Although we often fail in these departments, we must strive to practice them. Practice makes the master. We each are responsible for being the best person and professional we can be." Engineering manager, TransCore.

Chapter 13   Leadership
- "I'm most concerned about being able to recognize, adapt to, and overcome changing times and challenges." Engineer at major Midwestern engineering company.
- "I need to develop my leadership skills, be willing to take on more serious responsibility without self-doubt." Engineer at a major Midwestern engineering company.
- "Early in my career, a more senior gentleman at our company told me: 'This company focuses too much on doing things right and not doing the right things.' I [now] understand better what he meant. The optimum is to do the right things the right way." Chemical engineer at AIChE seminar
- "The challenge is how to manage time—time to work, time to relax." Junior engineer at ASME seminar.
- I'm concerned with the need for time management in the face of increasing work demands and reduced support." Chemical engineer at AIChE seminar

Chapter 14   Workplace
- "As a new and young engineer, I think the hardest part is learning

how to adapt to a world that is predominantly much older than you." Chemical engineer at AIChE seminar.

- "I'm concerned about office politics. What do you do when something 'serious' arises and it seems as though nothing can be done or said to help the situation?" Engineer at Tau Beta Pi seminar.

Chapter 15   Dealing with Stress and Having Fun

- "I'm most concerned about the great unknown. I have spent my entire life (I'm 25 years old) in an academic environment of some kind. So the real world is a bit of a mystery. I have concerns about my own skills relative to those with maybe a bit more hands-on experience. I have concerns about my ability to function effectively in large groups. I work well in groups up to about 25 people or so. In large groups, however, I have a tendency to withdraw and recede. Until now, I don't think that this has been a severe detriment, but I have concerns about that part of the real world, and how I would function in it." Engineering student at Tau Beta Pi seminar.

- "I don't know if I'm prepared for the 'real world.' It's easy to be afraid of the unknown." Engineering student at Tau Beta Pi seminar.

- "I need to keep stress levels lower by understanding possible outcomes." Engineering student, Webb Institute.

- "I'm concerned about not having enough time to be successful in the real world, have a family and have fun. I'm afraid that I will let people down when they depend on me." Chemical engineer at AIChE seminar.

- "I'm most concerned about expectations—what my company expects from me, what I expect from me, what my family expects from me." Engineer at a major engineering company.

- "I'm concerned that it's all downhill after college and I'll be bored with my job and engineering." Engineer at Tau Beta Pi seminar.

- "Your son saying 'engineers are the only people who need to be *told* to have fun' is *exactly* true. Completely dead on. That was probably my favorite part of the workshop. It opened my eyes to explaining the way engineers behave. My dad, who is an engineer, is a complete workaholic. Having fun seems like it is a chore for him. Your son's description is so accurate it's not even funny. Make sure you always include that in your presentation." Senior chemical engineering student at Cooper Union.

- "It's most helpful to hear someone telling me how to have fun. I'm a workaholic." Engineer at SWE seminar.

Chapter 16   Taking Action and Summing Up
- "I need to walk the walk. Stop talking and just do it." Engineer at a major Midwestern engineering company.
- "Fear stops us from doing many things, often for many years. Yet, we have to act in the face of fear and when we do we often find out it's easier than we thought. So fear is in our heads. Taking action gets us out of our heads." Engineering manager, TransCore.
- "Significant (effective) action can occur in very little time. It could be a three-minute conversation, a quick e-mail, a phone call, and so on. Epiphanies (transformations) can occur in literally instants and overcome years of being stopped (often by fear)." Engineering manager, TransCore.
- "I'm concerned that I won't be able to make a difference. There are so many injustices I see, and sometimes it seems as if one engineer can just get lost in the shuffle and ignore the things that he or she doesn't have to face on a daily basis." Engineer at Tau Beta Pi seminar.
- "I have concerns about the real world: Figuring out what I really want to do with my life; is this as good as it gets, or is something better out there?" Chemical engineer at AIChE seminar.
- "It forced me to think about things I don't usually think about— what's important to me, and so on." Engineer at ASME seminar.
- "I think learning these soft skills will be very helpful in my life. I want to really work hard to apply the principles put forth." Senior Cornell engineering student at SWE seminar.
- "Hearing about this stuff helped me identify certain critical aspects of my life. For a few years, I have been telling myself 'I'll do it when I have time' too much. I have not been on a real vacation in nearly three years, so that tells you a little bit of what I mean. I now realize that @#$! always happens, so it's up to me to make sure that I *make* time for myself. Similarly, I have to *decide* what will make me fully happy; what I want to be doing 5, 10, 20 years from now and start planning/living my life so that I can accomplish what I want." Engineer at Tau Beta Pi seminar.
- "I need to understand how to be responsible for myself and my life." Engineer at major Midwestern engineering company
- "Reminder of non-technical skills of engineering. Reality check

to myself that I need to work on certain things." Engineering Trainee, Port Authority of New York and New Jersey.

- "I've lived in the 'real world' most of my life. Take what you get and deal with it as it comes. Make the best of it." Engineer at SWE seminar.

- "I feel it is very important to be exposed to this stuff early in one's career. Thank you!" Engineering Trainee, Port Authority of New York and New Jersey.

- "I am a firm believer that work is meant to be an activity of self-fulfillment and enjoyment. Most young people like me are in the phase of shaping who they are. During the first 10 years of our professional life, we start the process of self-realization, beginning to develop what we like and dislike. Theories told are only helpful when practiced. It is a discipline, and [learning about soft skills] reaffirms this discipline. It reinforces the ideas and also allows us to reflect upon them." Electrical engineer at Bovis LendLease.

- "I heard you speak a few years ago when you were at our college, and now that I've been working for a couple of years, your presentation means something different than what I got out of it before. Thank you!" Engineer at SWE seminar.

- "I'm concerned about the path that I am on. I am hard working and know that I have a successful future ahead of me. I've got plans to go to grad school and I want to do research and make a difference in someone's life. I'm scared that I'm going to spend so much time and effort working that I won't have a chance or will forget to take time to settle down, get married, and have kids. So, mostly I'm scared about entering a working world that isn't going to allow me to have the family life I want as well." Engineering student at Tau Beta Pi seminar.

- "I have learned that the outside world actually craves a well-rounded engineer—an engineer who knows how to manage his or her time and participate actively at the place of work and, while doing all this, can have some fun." Chemical engineer at AIChE seminar.

- "Thanks for reminding me of what is important in life." Engineer at a major Midwestern engineering company.

- "The best way to learn to swim is to be thrown into the pool under the proper supervision, after having had a number of swimming lessons, of course. By and large, I gave low scores to my "young engineers" because they are young. Most of these skills come from experience, something they do not have *yet*. In general, I think the "young engineers" I have known will score well given time and opportunity." Engineering Manager, Port Authority of New York and New Jersey.

- "Laying out all these ideas, saying something about them and bringing them out into the open and being honest and frank about the 'real world' is good. Hopefully, some of these people will understand what this is all about because these are my peers and I'm going to be working with them for my entire life." Engineering student, Binghamton University.
- "I'm concerned that I will be comfortable with my skills/weaknesses, know my limits, and know how to work my best every day." Engineering student, Binghamton University.
- "We need to talk more about fears . . . everyone has them. Make it clear not to be afraid to ask questions or to state your opinions in the real world." Engineering student, Binghamton University.
- "But all of this is easier said than done." Engineer at Tau Beta Pi seminar.

3. LIST OF PROFESSIONAL SOCIETIES

The importance of professional engineering societies to your career and professional development cannot be understated. Membership in one or more societies in your field of interest allows you to keep current with technical developments in the profession through publications and attending conferences, and affords you opportunities to get involved in committees at local, regional, and national levels. Below is a brief alphabetic list of some of the major engineering professional societies and their Web sites that you may wish to contact for more information, if you are not already a member. You should talk to colleagues about what professional societies they are involved in and learn how they are getting the most value out of their association.

American Institute of Chemical Engineers (AIChE)—
    www.aiche.org
American Society of Civil Engineers (ASCE)—www.asce.org

American Society for Engineering Education (ASEE)—
www.asee.org
American Society of Heating, Refrigeration and Air-Conditioning
Engineers (ASHRAE)—www.ashrae.org
American Society of Mechanical Engineers (ASME)—
www.asme.org
Institute of Electrical and Electronics Engineers (IEEE)—
www.ieee.org
Institute of Transportation Engineers (ITE)—www.ite.org
International Federation of Automotive Engineering (FISITA)—
www.fisita.com
National Society of Black Engineers (NSBE)—www.nsbe.org
National Society of Professional Engineers (NSPE)—
www.nspe.org
Society of Manufacturing Engineers (SME)—www.sme.org
Society of Women Engineers (SWE)—www.swe.org
Women's Transportation Seminar (WTS)—www.wtsnational.org

Note that the Cornell Library Engineering Research Guide for
Engineering Societies and Organizations Web site (www.englib.cor-
nell.edu/erg/soc.php) lists societies by name and by acronym. The
International Directory of Engineering Societies and Related Organizations, Engr Diretories TA 1 E37, published by the American Society of Engineering Societies, lists over 1300 engineering organizations and includes their web/email addresses, description, list of
publications, conferences, officers, and dues.

INDEX

Stuff You Don't Learn in Engineering School. By Carl Selinger.
ISBN 0-471-65576-7 © 2004 the Institute of Electrical and Electronics Engineers, Inc.

ABOUT THE AUTHOR

Carl Selinger is Principal of Carl Selinger Services, a private consultancy that helps aviation and transportation organizations develop business strategy and apply new technologies. With more than 35 years experience in business, government, and academia, he finished a 30-year career with The Port Authority of New York and New Jersey, leaving in 1999 as Manager of Business Development for Aviation. In that position, he developed business, concessions, and technology initiatives to improve services and increase revenues at Kennedy International, LaGuardia and Newark Liberty airports. His motto is: "Ideas are always welcomed!"

Carl Selinger's career embraces a diverse set of activities including: wireless applications at airports, airport operations software, airport advertising, prepaid phonecards, CNN Airport Network, public Internet stations, car rentals, and interactive multimedia web-based payphones. Past projects included: developing preconditioned air systems at aircraft gates; managing 5,000 payphones at all Port Authority facilities; marketing Pan Am's $500 million major aircraft maintenance base at JFK; planning the reverse-flow express bus lane on the New Jersey approach to the Lincoln Tunnel—now 30 years old; managing the Manhattan Staggered Work Hours Program; authoring the Port Authority's first "Strategic Plan"; and guiding the start-up of the successful airport shuttle between Ridgewood, New Jersey and Newark Liberty Airport.

Carl has taught aviation and transportation courses for more 25 years. He now teaches courses at The Cooper Union, the State University of New York Maritime College's School of International Transportation Management, and at the Aviation Institute of the City University of New York's York College. He is a faculty advisor to Cooper Union's student chapter of Tau Beta Pi—the national engineering honorary society, which elected him "Eminent Engineer"

Stuff You Don't Learn in Engineering School. By Carl Selinger.   **177**
ISBN 0-471-65576-7 © 2004 the Institute of Electrical and Electronics Engineers, Inc.

and "Most Active Alumnus." Mr. Selinger is very active in Cooper Union alumni activities, was honored as "1993 Alumnus of the Year," is now president of the 11,000-member Alumni Association and serves on the Cooper Union Board of Trustees.

Selinger holds civil and transportation engineering degrees from The Cooper Union, Yale University, and Polytechnic University. He is very active professionally in Airports Council International (ACI) as liaison between committees on Business Information Technologies and Commercial & Business Management, the American Society for Engineering Education (ASEE), the American Society of Civil Engineers (ASCE), the Institute of Transportation Engineers (ITE), and the Women's Transportation Seminar (WTS), which honored Carl as the Greater New York Chapter's "1998 Member of the Year," and where he mentors young professionals. Carl is Honorary Member of the Sperry Transportation Board of Award, and has chaired the ASCE's National Transportation Policy Committee. He now serves on the Advisory Group developing an Engineering Management Certification for ASME and the founding engineering societies.

Among his diverse professional interests, Selinger speaks frequently to engineering groups, and facilitates focus groups and "Synectics" brainstorming sessions. His unique seminar—Stuff You Don't Learn in Engineering School—has been presented to more than 1000 young engineers during the last decade at engineering organizations, engineering schools, and professional society meetings. Carl's goal is to help engineers learn the nontechnical "soft" skills that are important to be more effective and happier. Carl's articles on these topics appear monthly in the Careers section of *IEEE Spectrum* magazine, where he is Contributing Editor.